COLD WAR
TEST PILOT

Dedication

This book is dedicated to the test aircrews, scientists, engineers and technicians with whom I had the privilege and honour to work while serving at the Aeroplane and Armament Experimental Establishment, Boscombe Down, between 1970 and 1988.

Photographs

Photographs are from the author's own collection and by kind permission of Photographic Division, Boscombe Down (A&AEE) or as otherwise attributed. Additional photographic material is from public-domain sources such as the British Library photographic archive. All practical efforts have been made to ensure that copyrights have been respected.

COLD WAR
TEST PILOT

Flying and testing military fast-jets
of the era – and how I survived
to tell the tale

a memoir
by
Group Captain

Ron Burrows

AIR WORLD

AIR WORLD

COLD WAR TEST PILOT

First published in Great Britain in 2021 by
Air World
An imprint of
Pen & Sword Books Ltd
Yorkshire – Philadelphia

ISBN 978 1 39909 070 4

Typeset by SJmagic DESIGN SERVICES, India.

Printed and bound by CPI Group (UK) Ltd, Croydon, CR0 4YY

Pen & Sword Books Limited incorporates the imprints of Atlas, Archaeology, Aviation, Discovery, Family History, Fiction, History, Maritime, Military, Military Classics, Politics, Select, Transport, True Crime, Air World, Frontline Publishing, Leo Cooper, Remember When, Seaforth Publishing, The Praetorian Press, Wharncliffe Local History, Wharncliffe Transport, Wharncliffe True Crime and White Owl.

For a complete list of Pen & Sword titles please contact

PEN & SWORD BOOKS LIMITED
47 Church Street, Barnsley, South Yorkshire, S70 2AS, England
E-mail: enquiries@pen-and-sword.co.uk
Website: www.pen-and-sword.co.uk

Or
PEN AND SWORD BOOKS
1950 Lawrence Rd, Havertown, PA 19083, USA
E-mail: Uspen-and-sword@casematepublishers.com
Website: www.penandswordbooks.com

Contents

List of Plates

About the Author

The author's first operational flying tour was on No 43 (Fighter) Squadron flying Hunter fighter/ground attack aircraft on close air support missions in the Aden Protectorate. After two years of operations over the deserts and mountains of what is now Southern Yemen, he returned to the UK and qualified as a flying instructor, following this with two instructional tours, first on Oxford University Air Squadron and then on the staff of the Central Flying School at RAF Little Rissington. In 1969, he was selected for test pilot training at the US Navy Test Pilots' School, at Naval Air Station Patuxent River, in Maryland, graduating a year later.

While there, he flew thirteen different US aircraft types including F8K Crusader, F4J Phantom, F104 Starfighter, and A4B Skyhawk, and won the US Navy League Award as the outstanding member of his class. Posted then to the Aeroplane and Armament Experimental Establishment (A&AEE) at Boscombe Down to take up the first of his four test-flying tours, he became RAF project test pilot on the tri-national cockpit design team for the Multi-Role Combat Aircraft (MRCA), later to be named Tornado. During his subsequent six years at A&AEE, he flight tested all of the current and new fast-jets and trainers of the time, either newly in service or about to enter it. His projects included Phantom, Harrier, Jaguar, and Buccaneer, and he was the first RAF pilot to fly the MRCA, flying from the German flight test centre at Manching near Munich. He also carried out hot-weather trials on Buccaneer in Arizona, cold-weather trials on Jaguar in Alberta, and weapon and systems flight trials on Sea Vixen, most marques of Canberra, and even the Nimrod MR1, which as a fast-jet pilot was unusual. He later formed and led the Sea Vixen/Buccaneer/MRCA fast-jet test unit, working out of the old 'C' (Royal Navy) Squadron hangar at Boscombe, before merging his small but special team into 'A' (Fighter Test) Squadron, at the same time becoming its Senior Test Pilot. After completing a staff tour as a NATO flight-test programme manager in Munich and later graduating from the

National Defence College, he joined the defence operational requirements branch in the Ministry of Defence as a wing commander.

In 1980, he returned to Boscombe Down, again to 'A' (Fighter Test) Squadron, this time as its CO, once more to be engaged in flight testing all the new and current fast jets and trainers of the time. He was the CO of the squadron during the period of urgent flight trials that became necessary in support of the re-capture of the Falkland Islands following the Argentinian invasion. During this time, he flight-tested Harrier GR3 in new weapon configurations, conducted air-to-air refuelling trials with the improvised Vulcan tanker, and carried out take-off trials using the carrier-deck ski-jump. After a further staff tour as Assistant Director Flying (MOD Procurement Executive), overseeing military flight testing throughout the UK, he was promoted to group captain and posted back to Boscombe Down, this time as chief test pilot and Superintendent of the Test Flying and Test Training Division. He was awarded the Air Force Cross in 1976 and was elected a Fellow of the Royal Aeronautical Society in 1987.

Author's Foreword

Nothing I have done in my long and lucky professional life ever held quite such a sense of mission, nor was quite so varied and stimulating, as my flying career in the Royal Air Force. Putting aside the mercifully few years that I endured behind a desk in the Ministry of Defence, I was first and foremost an aviator; and I was given so many different flying tasks and roles to perform in so many different places in the UK and overseas that it seems now, looking back on it, that my feet hardly ever touched the ground. Every week held fresh challenges, every year something new. As I perhaps too often say: flying was better than working for a living! And even though I subsequently held some quite demanding senior positions in other professional fields in the thirty years that followed my RAF career, I have always been an aviator in my heart and in my head. The RAF helped to form me in my adolescence; it educated, trained, and coached me throughout my career and gave me the skills that led to the little triumphs in my life; it expanded my horizons to distant corners of the globe; and it gave me the opportunity to find and pursue the things that I was good at.

I never came much into the public eye, although I was once reluctantly the subject of a 3-page colour article in the *Sunday Telegraph* magazine (a RAF Public Relations Department initiative!), and later, as the chief test pilot at Boscombe Down, I popped up here and there in Brian Johnson's BBC 'Test Pilot' series (some of which may still be found on YouTube, by the way), and his book of the same title. I was just one of many military test pilots of the period working quietly behind the scenes alongside the scientists and engineers at MoD testing establishments such as Farnborough, Bedford, and Boscombe Down, each of us making our own individual contributions to British military aviation research, development, and operational evaluation.

Considering what has happened in the world during the decades since I left the RAF, I feel fortunate to have served when I did. Throughout my service career, the Soviet Union and its Warsaw Pact allies posed the principal and major threat to the security of the UK (and its NATO partners). Indeed, my retirement from the Service practically coincided with the collapse of the USSR, although I'm fairly certain that my departure from the scene won't have been the cause of it! Whatever we might think about that Cold War period now, the Soviet threat felt very clear and very present then. For the servicemen and servicewomen of the time, and for the civilian scientists and engineers of the Ministry of Defence with whom I worked closely, this was the threat that drove our thinking and directed our work in whatever roles or arm we served. Our motivation and purpose was clear: constantly to improve our technology and capabilities so as to counter the changing threats that the Warsaw Pact forces posed to our security; to deter aggression against us through demonstrating our ability to retaliate meaningfully; and to defend the nation and thereby mitigate the damage likely to be caused should attack actually come. Things were simpler in those days than in more recent times, and our objectives clearer and more understandable to the population at large. There was also a very real sense that we were the defenders of our shores and the guardians of our freedoms; and we felt proud to be part of that, as I am sure our successors do just as intensely today.

By the very nature of their work, all test pilots could probably describe their test flying careers as unique, but few of my generation will have enjoyed such wide and varied aviation experiences as I have – or trodden a path so littered with notable thrills, spills, and narrow escapes. This book will tell you a lot about all those memorable events and about the work of test pilots. It is a tale spanning my thirty years as a professional aviator – a flying career like any career, pulled this way and that by unpredictable fate and improbable luck. While not especially important or influential in the scale of things, I had a wonderfully varied, immensely fulfilling, and at times truly exciting flying career. This is a tale written not to boast, but more to reflect and remember, and I hope, to inform and entertain too; and to set it all down on paper as a sort of valedictory footnote of an important aviation era – the last thirty years of the Cold War. And even if you are not especially interested in aviation and tales of derring-do, I hope that it will anyway make good reading.

Ron Burrows

Chapter 1

On 18 January 1943, my father, Steve, then an RAF flight-sergeant, flight-engineer on the Avro Lancasters of No 44 (Rhodesia) Squadron*, obtained a five-day leave pass. In the preceding forty-eight hours, he had survived two long-range night-bombing missions over Berlin, bringing his total missions over Germany and Italy to thirty-eight since the previous July. He deserved a spot of leave.

Nine months later, exactly to the day by my subsequent calculation, I arrived into this world and took my first breath. It would appear, therefore, that I have No 44 (Rhodesia) Squadron's adjutant to thank, at least partly, for my existence – or at least, for the timing of my arrival. It was his timely signature on my father's leave pass (as well as the BSA motorbike on which my father made his mad dash back to Evesham to see his new wife on that fateful day) that was also to make me a Libran son born under the astrological 'Air' sign. If his leave had been granted just four days later, I would have been born a Scorpio, having 'Water' as its element. As we shall see, those four days and this elemental difference may have determined the whole course of my future life!

But I sense your eyebrow rising … so I hasten to assure you that I don't believe in that star-sign stuff either. However, it is uncanny, being that I was born on the cusp, how these two elemental influences, air and water, would vie with each other for the rest of my life.

My father left school at 14, put on a miner's helmet a day later, and followed his father into the coal-rich strata of a Northumberland colliery. It seemed the obvious thing for him to do. All the young men in his village did the same. He didn't endure it for long, however; the mining heritage

---

* No 44 (Rhodesia) Squadron was re-titled as such in honour of the Rhodesian contribution to Britain's war effort. It also recognised that up to 25 per cent of the air and ground crew of the squadron were Rhodesian.

in his genes was not strong enough to overcome his dislike of hard labour and coal dust, and he was soon standing in the queue at Newcastle's RAF enlistment office. By 1937, he had signed up to train as an aircraft engine mechanic and was soon fixing the engines of Sunderland flying boats at RAF Calshot, and those of Hampdens and Liberators at RAF Honeybourne. In April 1942, he volunteered for aircrew duties and qualified as a leading aircraftman air-gunner at RAF Walney Island on Defiants. Only a month later, on promotion to the rank of sergeant, he moved to RAF Waddington to train as an Avro Lancaster flight-engineer.

The Lancaster Mk B1 cockpit, like most British bombers of the time, was configured for single-pilot operation and was thus equipped with only one set of flying controls. The flight engineer sat on the pilot's right-hand side on a 'Dicky' seat, a seat that could be folded out of the way to allow the bomb-aimer to go forward to his position in the aircraft's nose – as well as to clear the way for crew escape. It was from this 'Dicky' seat that my father managed the aircraft's four Rolls Royce Merlin XX engines and all the aircraft's other engineering systems, such as those for fuel, hydraulics, and electrics. Each of the four supercharged Merlins delivered 1,240 horse-power up to 10,000ft, and maintained over 1,100 horse-power even up to 18,000ft despite the reducing ambient air pressure. From the control panels of his flight-engineer station, Steve was controlling the power equivalent to that required by a medium-sized English village of the time.

In addition to this weighty role, he was also expected to fly the aircraft back to friendly territory if the pilot was wounded during flight. His crew's survival, or at least their avoidance of imprisonment in a German POW camp, would have depended upon it. Pilot incapacitation during operations was more than just a remote possibility – Steve's log book lists several instances of damage to his aircraft from anti-aircraft artillery fire or from strafing attacks by enemy fighters, so it was wise to be prepared.

His preparation for this contingency included basic pilot training on the twin-engine Airspeed Oxford and some instrument flying practice in the Link Trainer. This did not amount to a lot of time at the controls, and it was nothing like a full pilot's training course – but it was enough for the purpose. And with a little help from his navigator, he should have gained sufficient flying skill to pilot his Lancaster to a safe bale-out area if called upon to take over. Luckily, the young Steve never had to perform this role, but if he had, knowing him as I eventually did, he might well have been tempted to try a landing too. He would almost certainly have flown the aircraft from the pilot's seat from time to time – at a safe height in the cruise for practice, or if the pilot had to go aft for any reason – but it is inconceivable in an aircraft with just one set of flying controls that he would ever have been allowed to try a landing.

Anyone who has flown a heavy, propeller-engine, tail-dragger will know that more than a little skill is required to land such an aircraft safely – skills certainly beyond the abilities of anyone with just a few hours of basic flight training. Accident records are littered with instances where even experienced pilots of such aircraft have lost control during take-off or landing due to the huge 'swing' forces present, forces that can lead to a violent, uncontrolled rotation of the aircraft in the horizontal plane (a so-called 'ground loop') if the pilot's concentration lapses, even for a second, at these critical times. To get technical for a moment, 'swing' and 'ground-looping' phenomena in tail-wheel, propeller aircraft are due to several factors conspiring. Firstly, the centre of gravity lies aft of the main-wheels, which is destabilising (especially with a freely castoring tail-wheel); secondly, the spiralling prop-wash playing on the fin or fins creates strong yawing forces that vary with power and airspeed; next, the combined engine torque of the engines (all four rotating in the same direction on the Lancaster) puts more weight on the right wheel than on the left (thus creating a yawing moment due to higher rolling friction); and finally, the gyroscopic precession forces acting on the rotating propellers produce sudden yawing moments when the aircraft's pitch attitude is altered (for example, when the tail is raised or lowered during take-off and landing respectively). If the airspeed is high enough during a ground loop, moreover, the forward-going wing will develop significantly more lift than the other, causing it to rise skywards. This is the point at which the rear-going wing's undercarriage is liable to be torn from its mountings as its wing-tip begins to plough a furrow in the airfield's nicely-mown grass. In the extreme, this could end up most inelegantly with the aircraft lying on its back with what is left of its undercarriage pointing helplessly upwards.

I once watched a Sea Fury do exactly that just after touch-down on Boscombe Down's main runway, and end up on its back within seconds. The pilot was lucky to survive his half-hour of incarceration, trapped in his inverted cockpit while his fire-service rescuers freed him, and while high-octane aviation gasoline dripped from the aircraft's fuel vents onto scorched metal.

My father no doubt appreciated this danger. Nevertheless, he told me that he'd have felt compelled to have a go at landing if a crew member had been unable to bale out, say, through injury. 'I couldn't have just baled out, could I?' he told me. 'What? And leave one of my crew to go down with the aircraft? No way!'

After all, he'd been responsible for controlling the engines while the pilot carried out hundreds of landings so he'd have known what a good approach and flare should look like. He was also a plucky little chap and I'm sure he would have given it a go if he felt he'd had to. With his training, he would probably have got the main-wheels down onto the ground in roughly the right place too. But as soon as he throttled back and the tail went down, some of those malign swing forces would have kicked in with a vengeance. And being a bit of a short-arse, his little legs might not have been long enough to check the ensuing excursion. Once he'd allowed a swing to develop, it would have been game-over – the inertial forces would have developed so quickly that it just couldn't have been stopped. The aircraft would have then almost certainly slewed out of control, or at best, set off on a safari into the long grass – probably ending up on its belly, minus a wheel or two. Even so, his injured crew might have limped away from the ensuing smoking heap, which might have been a whole lot better than the alternative.

By the spring of 1943, my father had flown over 300 hours, surviving thirty-six long-range night-time operational sorties. In his second tour, after gaining his commission and a brief respite from operations as an instructor at No 1654 Operational Conversion Unit at RAF Wigsley, he was already back on operations again by the August of that year. He flew a further twenty operational sorties over Germany and France (the latter in support of the Normandy landings), and had logged a further 370 flying hours before the end of the war. He was awarded the Distinguished Flying Cross on 12 February 1944 and, in May 1945, he became the squadron's flight-engineer leader on promotion to the rank of Squadron Leader. He was then just 26 years old.

Chapter 2

I joined the Royal Air Force just seventeen years later in April 1962, entering Initial Officer Training at RAF South Cerney as an officer cadet on a direct-entry 'supplementary list' commission. I'd signed up to serve until the age of 38. My father hadn't encouraged me in this at all; in fact he was dead against the idea when I first declared my interest, thinking that to enter the RAF on short-service terms would limit my career – which, all things being equal, it could well have done. The minimum entry requirement for this type of short-service entry was (surprisingly) only four GCE 'O' levels including mathematics and English, which I had miraculously managed to exceed (by one) by staying on an extra year at school. Brighter entrants with two or three 'A' levels would have carried out their officer and flight-training at the RAF College at Cranwell, and would have been groomed from the start with expectations of promotion to the upper echelons of the Service. The career expectations of us direct-entry officer cadets were more modest, but for me, career and promotion were the last things on my mind. I just wanted to fly.

But this had not been my initial career choice. With no better idea of what to do with myself after leaving school, I had joined the Midlands Electricity Board as a student-apprentice electrical engineer, studying for a Higher National Certificate. In the gloom of that first winter, however, bent over my books in a depressing pool of anglepoise light, I came to realise that this was not likely to be enough for me. I wanted there to be some adventure in my life, some travel, something really interesting to do and to talk about. Electricity distribution in the Midlands didn't seem to offer much of that – not for me at any rate. Nor can I say that the prospect of five years of study enthralled me either. No one would have described me as the studious sort.

It was in this listless mood that my eye was drawn to a half-page RAF aircrew recruitment advertisement in a Sunday newspaper colour supplement.

Above some buoyant narrative that described a life of adventure and travel, there was a picture of five young men wearing sky-blue flying suits and yellow 'Mae-west' life-jackets walking towards the camera. It seemed they were engaged in animated and happy banter, while in the background, a gleaming white Vulcan bomber basked splendidly in the sunshine. The image captured my interest immediately. 'Hmm, that's a bit more like it!' I thought at once. The photograph encapsulated the very essence of what my subconscious mind had sought. It was like a light had been switched on and my pulse began to race. Until that moment, I had never dreamt that such a career might be open to me; but reading on, I was astonished to find that I might actually meet the minimum entry requirements. I remember the rising surge of excitement that I felt as my eyes raced along the lines twice-over to make sure that I had not miss-read it.

In those days, there were vocational training schemes in almost every kind of occupation. All the military services, the civil service, the merchant marine, the water and energy providers, the banks, and many large companies ran their own training schemes – some even their own training schools and colleges. Indeed, the great majority of school-leavers wanting to build a career would follow this route into professional employment. There were far fewer university places available anyway. Only about ten per cent of school-leavers went to university then, and so training schemes were the norm rather than the exception. In fact, for most employers, including the RAF and the other armed services at the time, it was the preferred entry route, with 'A'-Level and 'O'-Level entries available to suitable candidates. There were similar training schemes in law, banking, accountancy, engineering, and architecture, to name just a few – all structured as a 'sandwich' of part-time study combined with work-experience – learning from and working alongside those already qualified in the field. To my mind, it's difficult to think of a better way of doing it – at least for those who already know what they want to do (or be) in life after school.

In my earlier school days, as I had begun to turn my thoughts to what I might do for a career, I had at first been smitten with the idea of going to sea (the pull of Scorpio's elemental 'water' temporarily in the ascendant perhaps, rather than Libra's 'air'?). Many of the shipping lines offered vocational training schemes, and for a while I had become besotted with the idea of joining Shell Petroleum as a student deck (navigation) officer. Shell's scheme was a five-year sandwich course mixing study ashore with time at sea, with a subsequent career path open all the way up to captaincy. Like the RAF's direct-entry scheme, theirs was another requiring only four GCE

'O' levels for entry; and both Shell and BP were then recruiting actively. The image of myself as officer-of-the-watch dressed in crisply-pressed whites, manoeuvring my ship into some exotic tropical port had captured my boyhood imagination. But after a little friendly advice from a retired merchant seaman and some cool reflection on the likely reality – months afloat staring at empty horizons with moorings at remote refineries – the prospect soon lost some of its attraction.

But now, as I read and re-read this colourful RAF recruitment ad, with the hairs on the back of my neck standing on end, my career path ahead suddenly seemed absolutely clear. The elemental 'Air' had won out! Why had I not thought of this before, I wondered?

I handed in my notice to the MEB on the same day that I sent in my application for RAF pilot training, never for one moment thinking that I would fail to be selected. What over-confidence; what naivety, you might say! I was successful nevertheless, and it is a mark of the times, or perhaps a reflection of the urgency of RAF recruitment at that time, that I would start initial officer training only ten weeks later. And that short time included the three days of tests, interviews, and doing silly things with 55-gallon oil-drums and wooden planks at RAF Hornchurch, then the centre for officer and aircrew selection.

My initial officer training lasted four months between April and August 1962. It was a busy, fast-paced course which included service studies, a lot of physical training, a similar amount of square-bashing, and several route-marches and escape and evasion exercises in the Brecon Beacons (with some more silly things to do, this time with scaffolding poles and coils of rope). I passed out (not literally) as a brand new acting pilot officer (probationary) in August and was sent to No 3 Flying Training School at RAF Leeming, Yorkshire, to join twenty-five other cadet-pilot trainees. On 3 September, not yet 19 years old, probably the youngest member of my course, and never having set foot in an aircraft in my life nor yet licensed to drive a motor vehicle, I took to the air for the first time in a Jet Provost Mk3. To reach the start of my pilot training had taken only seven months from first thinking about it to my first flight.

In those days, the RAF trained its new pilots on jets from the outset without any precursor training on simpler propeller aircraft. This approach would soon change however, due to the high failure rate experienced and the associated wasted costs. A few years later, pre-jet elementary training on the venerable De Havilland Chipmunk was introduced, not only to provide a kinder and perhaps more appropriate introduction to piloting, but also to filter out those who did not have quite enough aptitude. Nevertheless, the Jet Provost was good at its job for those who did.

I took to flying like a duck to flight (adapting the usual analogy), discovering that I had a real talent for something after all. Weekdays were for flying and ground-school, evenings were for study and preparation, but the weekends were for outings to the Yorkshire Moors and exploring the local hostelry. It was a work-hard, play-hard culture, and I grew up rapidly during that first flying year, quickly overcoming my initial aversion to the taste of beer to lubricate my chat-up lines on some pretty Yorkshire barmaids. This was where I fell in love for the first time (and the second and third!) – but marrying a Yorkshire lass was never meant to be. This was also a time before breathalysers and speed cameras, and so no one seemed to worry too much about having a few pints before driving home.

Speaking of driving, my first brush with death (the first of several as you will read if you stay with me long enough on this journey) was as a passenger in my friend John Proctor's Triumph TR3, returning from a night out in Ripon. I don't remember either of us having had a lot to drink, but on encountering a patch of black ice on the A1 (then still a single-carriageway), he lost control of his lovely vehicle, which eventually came to rest upside-down in a roadside culvert following a few hair-raising pirouettes. As the car's detritus fell about our ears, I remember feeling grateful that his hard-top had been fitted for the winter, for without it (and having no seat straps either in those days) we may well have come off an awful lot worse. Even so, it was an amazing feat of youthful contortion to extricate ourselves through the badly distorted window openings. Indeed, once out and able to shine a torch on the heap of crumpled, hissing metal from which we had just crawled, it was hard to see how we had managed to get out of it at all.

My flying instructor throughout the twelve-month basic flying course was Peter Isherwood, and I have a lot to thank him for. He was the perfect role-model for me at an important time in my adolescent development. The early disciplines and attitudes that he inculcated in me, his youngest student, helped me both to survive and eventually to excel as a military aviator. We must have spent a hundred hours together in his airborne classroom (as he called it), and hundreds more in the briefing room. Every one of those hours was a model lesson of a quality rarely equalled in my experience thereafter, and never surpassed. He once told me that I had the homing instinct of a bird, and tested it from time to time over eight-eighths cloud by asking me for a heading to steer back to base. More often than not, I could tell him the right direction out of pure instinct, without even thinking about it. In the days when navigation was a black art, unaided by GPS and electronic navigation displays, this would prove a useful attribute. But after he had brought this apparent magnetic sensibility to my attention, it proved fragile. If I tried mentally to reason where base must be with any kind of conscious logic, instinct failed me. I'm sure that we have all inherited some innate bird-like sense of orientation from our pre-homo-sapiens animal ancestors, although it must be such a slight sense that conscious thought will completely overwhelm it. Sometimes it is better not to think about things too much (and let the Force be with you!).

My favourite phases of the 160-hour flying course, perhaps needless for me to say considering what you already know from my brief résumé, were low-level navigation, formation flying and aerobatics, which thrilled and excited me beyond measure. These challenging, fast-moving, and spatial tasks would stretch any new pilot's pure flying skills and aptitudes to the limit. To be able to execute these tasks with a reasonable degree of competence, moreover, gave me a tremendous feeling of satisfaction, not to say outright pleasure. It was the thrill of all of this cavorting around the sky that would draw me on to becoming in turn a fighter pilot, a flying instructor, and a test pilot. Looking back over my subsequent career, I now see that I could not have chosen a more suitable profession nor done anything else with half so much success. I was so lucky, therefore, that the opportunity was there for me. As I write this in 2020, there are far fewer training places available in a much-reduced RAF, and I probably wouldn't pass the entrance tests anyway.

Having started with twenty-five student pilots on our basic flying course at Leeming twelve months before, only thirteen of us were to graduate in July 1963, a pass rate of little more than fifty per cent. This was not untypical of the RAF's basic flying training schools of the early sixties following the introduction of straight-through jet-training on the Jet Provost, and this illustrates why preliminary flying training on simpler aircraft became necessary. These days, fewer than thirty per cent will fail during the whole three or four years of training required for new fast-jet pilots to reach operational status. Some of this improvement must be due to better initial screening and aptitude testing, but with competition keener for fewer places, the bar for entry will also undoubtedly be set higher. Of the thirteen students to graduate from my course, I was one of seven to be selected for advanced jet training, while the remaining six, destined to become transport or helicopter pilots, would go on to train on the twin-piston-engine Vickers Varsity at RAF Oakington or the Bell/Westland Sioux at RAF Ternhill.

My advanced jet training was to have taken place at RAF Valley in Anglesey on the new Folland Gnat, a diminutive but quite nippy little tandem-seat jet that sat almost low enough on its undercarriage to be mounted like a horse. However, when I and my fellow student pilots were due to start this training, neither the airbase nor the aircraft were quite ready for us. Somebody in the Air Ministry then had the bright idea that we should carry out our advanced training on the twin-engine Gloster Meteor instead, and we were even given the technical manuals and other training material to swot up. Fortunately, the idea was dropped when a more sensible senior officer intervened. The twin-engine jet had already earned the nick-name 'meat-box' because it was a potential killer on one engine and would have required too big a leap of skill for such novices as we then were to master. Nor was there anyway a suitable Meteor unit then in existence at which such training could take place. And so, not wanting to interrupt training momentum, the Air Ministry decided that we should be sent instead to the Royal Naval Air Station at Linton-on-Ouse near York, to train on the Vampire T-11, then used by the Royal Navy for advanced jet training. Linton was the advanced flying training school for naval fast-jet pilots destined to join Fleet Air Arm squadrons flying Sea Venom and Sea Vixen, both twin-boom aircraft of similar De Havilland heritage to the Vampire.

It was while awaiting the start of this delayed course that I was able to arrange a short secondment to the Royal Radar Establishment's flying unit at RAF Pershore. The unit flew a range of aircraft including Canberra, Hastings, Hermes, Devon, and Varsity as test-beds for new radar and radio equipment, and I was able to fly in all of these as second-pilot, sometimes flying several different types a day. Most of this flying was relatively unchallenging, but to describe me as second pilot in the authorisation sheets was a bit of an exaggeration when in truth I was only permitted to get my hands on the controls for the easy stuff, and certainly not anywhere near the ground. But I loved the variety, and as an ab-initio pilot qualified only on one aircraft type – and a basic trainer at that – I found it interesting to observe how the crews of large, more complex aircraft operated. It was at the RRE that I was first exposed to the nature and disciplines of trials flying – albeit of a rather mundane sort – and I came to admire the pilots who could hop from one aircraft type to another and fly them all competently (for the most part). I only mention this because it was at Pershore that the idea of test piloting first took root.

After this brief diversion, I arrived at Linton-on-Ouse in August 1963 and would spend the next six months living with the Royal Navy rather than at RAF Valley as had originally been intended. Our course was rather confusingly retitled 'No 2 (RAF Valley) Course' to identify us as 'crabs' rather than 'fish-heads' (not that it would have been difficult to spot the difference). After completing a two-week technical ground school on the Vampire, a basic conversion to type and an instrument rating, the applied phases of this six-month, seventy flying-hour advanced course, included high and low-level navigation, night flying, and formation flying and tail-chasing – a syllabus similar to the basic course, but faster! The course was designed not only to develop flying ability on this higher performance aircraft but also to stream pilots for the different operational jet-flying roles. Depending upon the aptitudes and abilities demonstrated in the several different training phases of the course, successful student pilots would expect subsequently to be selected for operational training either on jet-fighter or on jet-bomber types. Those selected for fighters would progress on to the Hawker Hunter or the English Electric Lightning; those for bombers, either to the Canberra or to any of the three V-Bombers: Vulcan, Victor, or Valiant (the latter, not destined to remain in service much longer). New pilots sent to the 'V' Force, as we called it then, would probably spend their entire first tours learning the ropes as co-pilots to more experienced captains. This would allow them to build experience on these big, heavy, complicated, multi-crew aircraft, before taking command themselves. We Lightning, Hunter, and Canberra types would become first pilots (i.e. pilots in command) as soon as we qualified.

The De Havilland Vampire

The twin-boom, single-engine De Havilland Vampire, with its plywood and balsa-wood fuselage, was originally a DH experimental jet-aircraft design put forward by the company as a possible replacement for WWII fighters such as Tempest and Typhoon. The design was accepted by the Air Ministry and eventually put into mass-production as a fighter/interceptor in 1944. After prototype and engine development, it eventually entered service in 1946, just too late to engage with the Luftwaffe (and just beaten into service the year before by the Meteor, which did). The later Vampire FB.5, equipped with four Hispano 20mm cannon and carrying up to eight 3-inch rockets and two 500lb bombs, was flown by nineteen RAF squadrons at its peak, and saw operational service during the counter-insurgency campaign in Malaya (late 1940s and early 1950s). The Vampire was also operated by some thirty other air forces worldwide. Indeed, the only Western powers not to use the type were Germany, the Netherlands, Spain and the USA. The aircraft remained in front-line RAF service until 1953, after which it was relegated to secondary roles such as the advanced pilot training that I was about to undertake. It was eventually retired from RAF service in 1965.

With a cockpit designed before ejection seats had even been conceived let alone manufactured in the UK, the pilot sat on his parachute pack and was restrained with straps within a metal seat-tray. (Martin-Baker Ltd demonstrated its first ejection seat in 1946.) Without an ejection seat, the cockpit was relatively roomy (if badly laid out by modern standards), but the chances of baling out of it in an emergency without ending up draped around the tailplane were slim. When ejection seats were eventually retro-fitted, they certainly improved the pilots' survival rate, but the sheer bulk of the seat, with its cartridge pack and rails, made the cockpit feel very cramped indeed. This was especially true for the two-seat training version, the T-11, especially when pilots wore life-jackets and rigid 'bone domes' (instead of the original leather helmets and flying goggles). The depth of the seat's parachute back-pack, moreover, pushed the pilots' torso and eye position a good six-inches further forward than it would otherwise have been, making the instrument panel feel myopically close. For the same reason, the pilot's thigh length became critical in deciding who would and who would not be permitted to fly the aircraft – pilots with thigh-lengths too long risked a knee-capping on their way out if the seat was used in anger.

The Vampire was powered by a 3,500lb-thrust Goblin turbojet developed for De Havilland by Frank Halford who based his design on the original turbojet engines pioneered by Frank Whittle during the war. Compared to Whittle designs, Halford's engine used a single rather than a double-sided compressor with the inlet at the front, and a 'straight through' layout with the combustion chambers exhausting straight onto the turbine. Whittle's designs used a 'reverse flow' layout that piped the hot air back to the middle of the engine in order to reduce the turbine entry temperature, a safeguard necessary to protect the lower-performance turbine materials available at the time. The Goblin had a centrifugal compressor, sixteen individual combustion chambers, a single-stage turbine, and it idled with a characteristic whine at F-sharp that could be heard a mile away by a deaf dog. Unlike piston engines, which respond almost instantaneously to throttle adjustment, jet engines are generally much slower to react, particularly when at low power (low RPM) settings. For example, if you slammed the throttle fully open from the idle-stop in the Vampire, you had to tap your fingers for a good ten to fifteen seconds before full power was eventually achieved. This is quite a long time if you think about it: imagine having to wait ten to fifteen seconds for a meaningful response from your motor-car engine after you put your foot down to overtake! But worse than that, the exponential shape of the thrust/rpm curve meant that very little additional thrust was developed in the first three-quarters of those anxious seconds. In other words: nothing would appear to happen for quite a while after opening the throttle, and then, just as you were beginning to think that the fire must have gone out, you got this monumental surge forward like being hit from behind by a tidal wave.

This slow-response characteristic, common in all turbo jet-engine aircraft, can catch out the unwary. It is the reason why, when performing a landing approach in a jet, it is important to have your engine running in the higher-RPM range where response to throttle movement is faster, and from where full power can be generated quickly if you need it – critical if you decide to abort a landing approach and go around for another try. For carrier landings, a fast response is especially important because of deck movement (pitch and heave in particular) and wind disturbances (especially sink) on short final. The higher the RPM in a turbo-jet engine, however, the higher the thrust; and for a given drag on a steady glide-path, more thrust means more speed – just what you may not want when aiming for an on-speed fixed-point touch-down (so as not to over-stress the landing gear or use more runway than available). Therefore, sleekly-shaped aircraft, designed to slip through the air with minimum drag most of the time, need a means of increasing drag for a powered landing approach. This extra drag is provided by landing flaps and slats on most aircraft, but on some aircraft, like the Buccaneer for example, air brakes are extended on approach as well.

All this is very well understood for a normal powered approach, but when practising glide approaches onto landing strips, as pilots of single-engine aircraft do quite often to simulate engine failure, it is generally done at idle power. And, as we have just heard, being at idle power in a jet all the way down to touch-down, puts the pilot exactly where he ought not to be as far as engine response is concerned, especially if at a late stage he decides to abort his approach.

Such were the circumstances in the first of my four aircraft accidents, the only one for which my piloting skills fell a little short of the mark!

On 6 September 1963 in Vampire XE852, I was briefed by my instructor for a general handling sortie, which was to include my first solo simulated engine failure followed by a practice forced landing at Holme-on-Spalding

Moor (abbreviated to 'Holme' because the full name was a bit of a mouthful). Holme, just north of the River Humber, was formerly an RAF bomber base in WWII but by the sixties it had become the testing airfield for the Buccaneer naval twin-jet light bomber made by Blackburn at their works in nearby Brough. My sortie included the usual general handling exercises – stalls, steep turns, and aerobatics, etcetera – and these were followed by a practice fire drill, which led into a simulated engine shut down (by closing the throttle to idle). All this took place at an altitude of about 10,000ft, which gave me a glide range of some twenty miles. Accordingly, I had arranged my flight to be close enough to Holme at that stage in my sortie to allow me to arrive overhead the airfield at about 3,500ft, which would give me the height required to manoeuvre into a position abeam my intended touchdown point at about 2,000ft. All this went swimmingly and according to plan. But when I turned onto the final approach at about 1,000ft, the wind-strength turned out to be higher than I had appreciated, and I soon found myself sinking below the ideal glide path.

Recognising this, I decided (sensibly, if a little too late) to abort the exercise and climb back up for another attempt, resolving to allow more height the next time around. But when I opened the throttle, the increase in thrust that I needed to arrest the descent simply failed to materialise in time. It was at this point that the long spool-up characteristic of jet engines really hit home (and it taught me to anticipate it better in future). My engine RPM began to increase, but ever-so painfully slowly; and for the next twelve excruciating bottom-clenching seconds before any useful additional thrust became apparent, the aircraft continued its seemingly inexorable descent. In such circumstances a pilot must resist the temptation to raise the nose, thinking perhaps that in doing so the gliding distance might be stretched. This is generally a recipe for disaster, a message hammered home again and again by our instructors. Giving in to the temptation would reduce airspeed below the minimum drag speed, and therefore cause drag to increase. This is not at all helpful in such circumstances because it not only reduces gliding range (contrary to the pilot's wishful thinking) but also puts him closer to wing-stall, loss of control, flick, crash, and burn, etc! Thus, despite that terrible sinking feeling as the ground rose up to meet me, I maintained the best gliding speed meticulously.

All this time, I could feel the thrust increasing gradually as the RPM needles inched their way slowly around the dial, and this little bit of extra thrust was certainly allowing me to reduce my glide angle. But it was not happening quickly enough. And it soon became obvious that I would make contact with the ground somewhat short of the runway. Under my oxygen mask I uttered a curse and gritted my teeth, realising what I might be in for. However, as the seconds ticked past, I was relieved to see that my touch down was likely to occur on the prepared undershoot area, provided for just the sort of inept misjudgement that I was now demonstrating. Although potentially embarrassing, I knew that landing just a little short of the runway on such a surface should not be catastrophic. The undershoot area was flat, firm, and clear of obstacles; and providing that I flared as usual and touched-down at the right airspeed, it would be little different from landing in the right place – or so I thought.

A few seconds later, just as I began to feel the Goblin's thrust press reassuringly into the small of my back, the Vampire's wheels touched down as smoothly and as gently as any normal landing, but about 100ft short of the runway threshold. Now, in normal circumstances, such a landing would have passed almost without consequences, except perhaps a ribbing from one's fellow students in the bar that evening and a bollocking by my instructor (if he found out). But unfortunately, Holme's runway had recently been re-surfaced, and the contractor had left a pronounced lip at the beginning of the new tarmac, which stood perhaps nine inches above the relatively smooth surface of the undershoot area. It was this lip that was to prove my undoing, and it led to near disaster. On striking the lip, my Vampire's nose-wheel buckled and collapsed, and I soon found myself careering along the runway at high speed with my nose grinding the ground and my twin tails pointing inelegantly skywards like the handles of a wheelbarrow. Even this might not have been terminal for the ageing aircraft, which might yet have been saved from the scrap heap. But the heat generated by the friction caused the plywood and balsa-wood fuselage to ignite.

It must have been quite alarming for the air traffic controller in his control tower to watch my Vampire streaking along his brand new runway like a wheelbarrow engulfed in flames, but I was unaware of the conflagration until the aircraft had ground itself to an undignified halt. It was only then that the smoke and flames caught up with me.

Modern ejection seats have a 'zero-zero' capability (zero height and zero airspeed), which means that a survivable ejection is possible even from an aircraft that is stationary on the ground – just the capability that might have been useful in the predicament in which I now found myself. But in those days, ejection seats needed a minimum of 200ft altitude (in level flight) and a forward airspeed of at least 90 knots to function as advertised, so there was nothing for it but to risk the flames. I unstrapped my harness, pulled down my visor to protect my face, and opened the canopy. The heat hit me immediately as flames licked around the open cockpit; and like a cat on hot bricks, I leapt over the side and made a run for it, fearing that the whole thing might explode if the flames reached the fuel tanks. Fortunately the fire service tender arrived at the scene within seconds and put out the fire, but the old girl would never fly again.

Well, of course, my instructor did get to find out, and I did receive a bollocking from the Group Captain station commander at Linton; but all-in-all, things could have turned out a lot worse and I had learned a valuable lesson. In future, practice forced landings would be carried out with half-flap (to increase drag) and increased engine RPM (to compensate) so as to make the Vampire's simulated glide representative of the real thing while also giving the pilot a faster engine response if he needed it. This technique was undoubtedly not new, but somehow it seems to have been left off the Vampire training syllabus in the hurry to set up our ad hoc course. In highlighting the potential dangers of idle-power descents by grinding my nose along Holme's brand-new runway, it seems therefore that I had unwittingly helped other junior Vampire pilots to avoid the same fate!

Chapter 3

The Hawker Hunter

Despite the above mentioned embarrassment, for which my hurt pride presses me to declare that I was not held wholly to blame, I successfully completed my Vampire course and was selected for fighter ground attack training on the Hawker Hunter at RAF Chivenor in North Devon.

The Hunter, first flown from Boscombe Down in 1951, was an elegant swept-wing design with pleasing lines, excellent handling qualities, and relatively high performance for the time. It was rated by most discerning pilots as a delight to fly. Although in Neville Duke's hands in 1953 it had 'broken the sound barrier' by attaining a level flight speed of 727.63mph, it was not truly a supersonic aircraft. Indeed, even without the drag of external weapons and fuel tanks, the production aircraft could only exceed sonic airspeed (Mach-1) in a dive, leaving parts of the aircraft still in subsonic airflow. To rate as a truly supersonic aircraft, the airflow over every substantial part of the airframe must exceed Mach-1, even around those tricky bends and junctions that it finds difficult to negotiate. But high subsonic speeds were quite high enough for the aircraft's role. Flying at or below 250ft above ground level, the normal low-level navigation airspeed would be 420 knots (484mph) with bursts at 540 knots (621mph) or even 600 knots (700mph) over hostile territory (or sometimes just for the sheer thrill of it!). These airspeeds were not arbitrarily chosen; they represented seven, nine, and ten nautical miles per minute respectively, which made it easier to navigate using stopwatch, compass, and map. These, by the way, were the only aids to low-level navigation available prior to the introduction of inertial navigation platforms and head-up displays, first fitted to the Harrier which entered service in 1969 (about which, more later).

It may also be worth mentioning that to cut and fold a map properly for use in the tight-fitting cockpit of a single-seat fighter was quite an art.

Our maps needed to be big enough to cover the route at a scale of 1:250,000 yet remain neat and manageable while galloping along in low-level turbulence, holding it with one hand while flying the aircraft with the other. Moreover, we would generally need two maps: one for the low-level route to and from an initial point (IP), a prominent feature close to the target area, and another larger scale map (1:50,000 or larger) for the run in from the IP to the target. Sometimes with photographs of the target area to carry too, and spare film-cassettes for the gun-sight camera, our cockpits and flying-suit pockets got a bit crowded. Good housekeeping, therefore, was a necessary skill if you didn't want your cockpit to become a slum!

I arrived at RAF Chivenor in March 1964 for a twelve-week Hunter day fighter ground attack (DFGA) operational conversion course. The course included about seventy flying hours in all, fifty of which were flown in the single-seat Hunter F6, the rest being flown with an instructor in the two-seat Hunter T7 for the demonstrations that preceded each phase. The great majority of the course, following a basic type-conversion and instrument rating, was spent learning the essential applied flying techniques required of DFGA pilots. These were: air-to-air and air-to-ground gunnery, air-to-ground (3' unguided) rocket firing, 1-v-1 and 2-v-2 air-to-air simulated aerial combat, and low-level navigation in up to 4-aircraft battle formations, which sometimes included a bounce by 'enemy fighters' somewhere along the route to test your mettle. 'Bounce' in this context meant being set upon by 'aggressor' aircraft flown by instructors who took much pleasure in scattering your formation to the four points of the compass and spoiling your carefully prepared plans.

In 1964, when RAF strength numbered about 150,000 personnel, there were eleven RAF Hunter squadrons operating from ten bases in the UK, Germany, the Arabian peninsula, and Hong Kong. To set this in context, the RAF inventory at that time included twenty-three different aircraft types (not including marques) and fifty-four operational airbases: three in Hong Kong; four in the Arabian Peninsula; two in Cyprus; five in West Germany; one each in Libya, Malta, and Gibraltar; and thirty-seven in the UK (not including staging posts in the Atlantic and Indian Ocean and other non-military airfields at which the RAF's university air squadrons were based). By comparison, at the time of writing (2019), RAF personnel strength is around 32,000 in total, and the service has no flying squadrons permanently based abroad except at RAF Akrotiri in Cyprus – although RAF squadrons and other flying units are deployed worldwide according to operational need and for training and logistic support purposes. The RAF now maintains only eighteen operational airbases worldwide, fifteen of these in the UK, one in Cyprus, one in Gibraltar, and one on Ascension Island.

The primary role of the RAF throughout my entire service life was the defence of Western Europe against attack by the Soviet Union. Consequently, the majority of the RAF's defensive and fighting assets during that period were deployed in the UK and West Germany. These included ground-to-air missile defences and early warning radar stations, medium and heavy bomber squadrons (carrying conventional and nuclear weapons), maritime attack and reconnaissance squadrons, all-weather air defence squadrons, and fighter/ground attack squadrons. When I first became operational in 1964, the RAF's fighter, bomber and ground attack aircraft inventory in the European theatre included Gloster Javelins and English Electric Lightnings for air defence; the English Electric Canberra and the 'V'-bomber fleet (Avro Vulcans and Handley Page Victors) in the medium and strategic bombing role; and the fighter/ground attack Hawker Hunters. Of the latter type, except for two close air support Hunter squadrons in the UK and two fighter reconnaissance (FR10) Hunter squadrons in Germany, the main focus of Hunter operations had by then shifted to the Middle and Far-East. In 1964, Hunter DFGA squadrons and fighter reconnaissance units were still active in Hong Kong (Tengha and Kai Taq) and in the Arabian Peninsula (Khormaksar in the Aden Protectorate and Muharraq in Bahrain). Thus, in 1964 when I graduated from my Hunter conversion course at Chivenor to become a rookie DFGA pilot, there was a very good possibility that my first operational posting would be in some far corner of the rapidly shrinking British Empire. Unfortunately for me, it would be the barren rocks of Aden rather than the exotic and colourful colony of Hong Kong that would become my home for the next two years or so. But it may be that providence intervened in selecting my direction of travel, for it was in Aden that I would meet Caroline, my future wife.

Into the Remote Places – Inscription of the crest of RAF Khormaksar. Aden.

I arrived in Aden in July 1964 as a newly made-up pilot officer to join No 43 (Fighter) Squadron at RAF Khormaksar, which was then the busiest airfield in the service. Not only did the station accommodate two Hunter DFGA squadrons (Nos 8 and 43) and a Hunter fighter reconnaissance flight (No 1417 Flight), it also accommodated two helicopter units operating Belvedere and Whirlwind; several short and medium-range transport squadrons operating Twin Pioneer, Beverley, and Argosy; and one Shackleton squadron employed on Army support and maritime patrol, supply-dropping, communication-relay, photo-reconnaissance, and ground-attack. Khormaksar was also the home of Aden Airways, the local airline which operated Vickers Viscounts and Douglas DC3s on routes around the Middle East and north-east Africa, and a staging post for RAF and commercial aircraft on their way to destinations further East.

At the time of my arrival, hostile incursions by north-Yemeni guerrilla groups into the Protectorate through the mountainous region around Jabal Radfan had begun and were already demanding a concerted counter-insurgency campaign. Over the next four years, until the British departure in 1968, guerrilla fighting in the mountains would be continuous, and terrorism in Aden itself would become more and more troublesome. By comparison with more recent and more savage middle-eastern wars the Aden campaign must now seem almost a foot note, but it was real enough for the participants and casualties on both sides. It was an 'old-fashioned' war in which, by and large, only the combatants were involved, battling with each other as if on some separate playing field removed from public view. Apart from the occasional incident in Crater and down-town Aden city, accompanied by curfews, security patrols, and roadside checks, civilian and off-duty military life had some semblance of normality – until perhaps towards the end of the occupation when things became a little more difficult, especially as the British began to pack up and leave.

The several distinct districts of the city were clustered in the crater and around the slopes of the long-extinct volcano of Jabal Shamsan, made famous by Pipe Major Alexander Mackellar's 'Barren Rocks of Aden', arranged while he was stationed there with the 78th Seaforth Highlanders. Until December 1963, it was a relatively peaceful place, with the Arab and Asian indigenous population rubbing along with the British government and service community in apparently harmonious interdependence.

Aden had originally been of interest to Britain early in the nineteenth century as an anti-piracy station to protect shipping in the Gulf of Aden and on routes to British India. After the opening of the Suez canal in 1869, it also served as an important coaling station. Then, following India's independence in 1947, when Aden became less important strategically to the UK, it still served as a major commercial port with refuelling facilities, and as a useful British foothold in southern Arabia and north-east Africa. There was a large BP oil refinery there, which generated much oil-related shipping activity. Liners navigating the Red Sea route would stage through the port en route to South Africa, India, the Far East and Australasia, and passengers of the Orient and Union Castle Lines would occasionally descend into the duty-free shops and bazaars of Steamer Point to strip the shelves of tax-free goods.

As far as living there was concerned, life as a member of HM forces was not uncomfortable. Most service and civilian families lived in the service houses and apartment blocks of Khormaksar and Ma'allah; and the single officers, NCOs, and other ranks lived in their respective messes on the base. There were two or three small hotels for visiting businessmen, and one or two eating places and bars to try out – if you were not too fussy about food hygiene and were desperate for a change from mess cooking. There were children's play groups and infant and junior schools, a service hospital and several beach clubs with swimming pools, bars and restaurants. If you wanted to sail or Scuba dive, there were clubs for those activities too around the harbour, with horse-riding and gliding also available from the nearby airfield of Sheik Othman, across the salt pans, a mile or two to the north.

There were also two English-language radio stations: the volunteer-run Aden Forces Broadcasting Association (AFBA), and the BBC's professional British Forces Broadcasting Service (BFBS). The local airline, Aden Airways, (a subsidiary of BOAC, now BA), served the principal airports of North and East Africa and Bahrain with their Vickers Viscounts, as well as the airstrips of the Hadramaut, Djibouti, and Somalia with their Douglas DC3s (more of which later). Most of this colonial lifestyle continued during the two years or so of my posting there, even as the tension increased and curfews and restrictions on movement began progressively to cramp our freedoms.

Despite some of the alleviating comforts just described, Aden was also a hot, dusty, and dirty place in which to work and live, a place permeated with characteristic tropical odours of bacteriological decomposition and inadequate drainage. Many southern-Yemeni people worked as menials and labourers, some as cleaners, some as ayahs for British families, but the brightest of them and some of the resident Asian population found employment as clerks, administrators, and middle-managers. They worked in the port and airport, in the refinery, in local government and municipal services, and at some of the British military bases too. And there was a small but entrepreneurial population of Arab and Indian businessmen that ran all the usual wholesale and retail enterprises that you might expect in such a conurbation. But there were also beggars in the streets and Arab families squatting in hidden corners of the city, living in utter poverty in shanty settlements, who eked out a hand-to-mouth existence picking up casual work where they could find it.

And sand got everywhere. It blew in with the wind and hung in the air, and infiltrated every nook and cranny of living space. Brand new cars, bought tax-free, soon rusted in the salty air and lost their gloss from the constant sand-blasting and a blistering sun. The tarmac got so hot during the day that, even motoring at slow urban speeds, road tyres squealed like a gangster's get-away at every roundabout. And while it hardly ever rained in Aden, the monsoon winds that blew across the Indian Ocean made the air so humid that that you could almost cut it with a knife. It was at these times of the year that I would sometimes ride my Honda-50 motor scooter around the airfield perimeter just to cool off; and night times without air-conditioning became a trial.

The guerrilla war into which I had stepped had its roots in the wave of Arab nationalism that had been sweeping across parts of the Middle East in the six years prior to my arrival on the scene. The political situation was complicated (as it still is today), and may therefore be worth an attempt to summarise as background to the operational flying that we DFGA pilots were about to undertake …

In 1958, Iraq had become a republic after a military coup ousted its monarchy. In that same year, Egypt's President Nasser formed a union with Syria to create the United Arab Republic (UAR), which was the precursor to building a larger pan-Arab state. Nasser almost certainly had ambitions for the UAR (with himself as its president) to increase Arab influence by absorbing other states into a federation over time. But this was not to be. The union with Egypt proved to be divisive within Syria's ruling hierarchy, and this rift came to a head in a successful coup d'etat in 1961, led by the disaffected Syrian military. The union with Egypt and the UAR consequently collapsed. Not to be deflected from his drive towards Arab nationalism, however, Nasser then turned his attention to the Yemen, where a republican coup was already underway against Saudi-backed Royalist Yemini forces. In October 1962, he sent 5,000 regular Egyptian troops into the Yemen, and by December of the same year, 10,000 more. By late 1963, the number had increased to around 36,000, and a year later, at about the time I arrived in Aden, to around 50,000. By late 1965, the Egyptian troop commitment to Yemen had reached a peak of 55,000 men, which included thirteen infantry regiments, a tank division, and several special-forces units and paratroop regiments. Yak-11 piston-engine fighters, MiG-15 and MiG-17 jet fighters, and Ilyushin IL-28 twin-engine bombers were also deployed, flown by hundreds of Egyptian pilots who had been trained in the Soviet Union. While not overtly declared as part of his plan, it is likely that Nasser's secondary objective was to rid the Aden Protectorate of the British. Russia backed Nasser, undoubtedly hoping to extend Communist influence in the region, and they supplied financial aid and materiel support. None of Nasser's forces were mobilised directly against the British, but their presence in Yemen fuelled and encouraged anti-British sentiment, emboldening the direct action of Arab activists who had motives of their own.

By 1963, these anti-British groups, with varying (and sometimes conflicting) political objectives, had begun to coalesce into two larger, rival organizations. These were the National Liberation Front (NLF) and the Front for the Liberation of Occupied South Yemen (FLOSY), who hated each other almost as much as they hated the British. In response to the growing unrest in the region, the British persuaded the fifteen (later sixteen) semi-independent states of the Federation of Arab Emirates that made up the Aden Protectorate to merge with the Colony of Aden to become a new, enlarged Federation of South Arabia (FSA). Within this new Federation, the British were able to influence and broker inter-relationships between the states in an attempt to improve stability through better cooperation in the region. Federal garrisons were also formed in up-country locations to police and protect the region. In 1964, the British Government announced that independence would be granted to the Federation in 1968, when the UK intended to quit the region for good. But this declaration appears to have satisfied neither FLOSY nor the NLF – these rival organizations regarding the Federation as British-inspired and thus illegitimate. From then on, NLF and FLOSY would not only continue to fight each other, but also engage the British and the Regular Federal troops of the FSA in a fight for ultimate control of the whole territory after the British had left. (I realise that I have risked losing you with all this background detail, but I hoped it might help to explain the mess that we had got ourselves into.)

Hostilities started in late 1963, just months before I joined No 43 Squadron, with an NLF grenade attack against the British High Commissioner at Khormaksar Airport. The grenade killed his adviser and a bystander (but not the High Commissioner himself), and injured fifty other people. On that day, a state of emergency was declared in Aden. The NLF and FLOSY then began a campaign against British forces in Aden, relying largely on grenade attacks – one such being carried out during a children's party in Khormaksar, killing a young girl and wounding four other children. Guerrilla attacks subsequently targeted off-duty British officers and policemen, with much of the violence carried out in Crater, the historic Arab quarter. Resident British land forces were increasingly unable to deal with the rising violence or stem the flow of weapons being smuggled into Aden through the Radfan mountains. In 1964, therefore, the British 24th Infantry Brigade was sent to Aden to scale up the British response. It was in support of these ground forces that the Hunters of Nos 8 and 43 Squadrons were deployed.

Gloria Finis – Glory is the End. No 43 (Fighter) Squadron motto.

The day following my arrival in Aden, after a tedious twelve-hour flight on a British United Bristol Britannia from Stanstead, I was scheduled for my first area familiarisation flight with the CO of the squadron, Squadron Leader Phil Champniss, in a two-seat Hunter T.7. New to the desert, it felt like I had stepped into an oven as we left our air-conditioned offices to walk out to our aircraft, which sat shimmering on the hot concrete in the heat of the day. The aircraft was almost too hot to touch as we checked it over before climbing into the cockpit and strapping ourselves in – and we started up the engine and got going as quickly as we could so as not to cook ourselves for too long. Sitting under the closed-down canopy of a T7 was like being in a green-house, and the sun would have baked us if we had not kept it cracked open like an air-scoop as we taxied fast to the take-off point. It was such a relief to feel the pressurised cooling air thundering into the cockpit as our wheels left the ground.

Leaving the runway behind and turning northwards, the contrast between the cool, green fields of my Hunter flying in North Devon and the barren scene that lay before me was stark. In my first view of the Protectorate from the T7's claustrophobic cockpit, the arid landscape was a mixture of empty scrubland and rolling sand, backed in the distance by craggy volcanic mountains that rose to 7,000ft and stretched unendingly like a barrier from East to West. Airborne sand and dust, trapped by the anticyclonic inversion, softened the horizon in every direction, blurring the boundary between sky and land – hazy-blue merging into hazy-brown so gradually that it was difficult to judge precisely if the aircraft's nose was pointing up or down. At latitude thirteen degrees north, the midday sun threw no shadows across the contours on that mid-summer day. The sandy ridges and undulations lost their contrast, and visual appreciation of the aircraft's height above the ground was seriously impaired.

I would come to know that it was not often as bad as this. Sometimes the air was crystal clear and you could see to infinity and beyond, but on this first flight and on many more like it to come, it was like flying inside a dirty goldfish bowl. There were also very few features to focus upon, and those that could be seen were of unfamiliar shape and size, so that distance was also sometimes difficult to estimate. My CO warned me that flying at low level and at high speed in such conditions would have its dangers, for rising ground could creep up unnoticed and bring things to a rather abrupt end. And if the ground was sometimes dangerously featureless, so was the general landscape.

On the coastal plains, there was little to navigate by. A few tracks and dried-up river courses broke up the endless scrub, but there were none of the features that one would normally depend upon – no towns, no railway lines, no woodlands, no rivers or lakes, and very few roads. And in the mountains, one craggy peak would look much like any other. The valleys between them meandered randomly in all directions with no particular orientation. I could see that it would take a bit of getting used to. But there was one feature that could generally be relied upon: the distinctive coastline of South Arabia where the sandy shore met the Indian Ocean, which was never far away for a fast jet. There were a few prominent features in the hinterland that I would also soon get to know (and love) – oddly-shaped rocky outcrops that stuck up through the sand like chocolate rock-cakes, the single main road running like a pencil-line northwards to Lahej, Dhala, and Thumayr, and the spectacular canyons and ancient walled-cities of the Hadhramaut.

To navigate at low level and reliably find a destination (or target) with such a paucity of features, especially at high speed, required practice, for there were no electronic navigation systems in those days that could offer assistance so close to the ground. The key to success was to plan meticulously and to mark-up routes in minutes and seconds, with headings for each track and pull-up points (for dive attacks) and expected fuel remaining at each turning point. Most of our flying was in 2-ship or 4-ship battle formations, and it was sometimes necessary to plan for the different tracks of widely separated wingmen, especially at turning points where they may need to cross over. We might also mark-in reminders for radio calls, for example for formation changes when starting the run-in for an attack, or before entering a narrow, steep-sided valley. When setting off on a route up-country, the stop-watch would generally be started at take-off, making allowances in planning for turning radius and acceleration to cruise speed. The clock would be re-set for each subsequent leg. Winds in southern Arabia were practically insignificant in comparison with the airspeeds we flew at, and it was usually not worth bothering to calculate their effect on drift or ground-speed. By flying planned headings and airspeeds accurately, turning points and targets would be reached within seconds of planned arrival times. To make it easier, routes were generally planned to use prominent features as turning points and for track and timing checks. Except when flying among the mountains, it was difficult to get lost. If all turned to worms, however, the best remedy would be to climb to an altitude from which some recognisable feature might be seen, re-orientate and then descend back to low-level to resume the route.

There was, of course, always a remote possibility that we might be set upon up-country by an opportunistic Egyptian fighter, so it was wise to keep eyes peeled, especially near the Yemen border. In retrospect, I begin to wonder why we did quite so much rushing around so close to the ground, except that it kept our skills honed for the European theatre.

When flying in northern Europe against Warsaw Pact forces equipped with squadrons of defending fighters, anti-aircraft guns, and surface-to-air missiles, it made sense to stay below the enemy radar and minimise exposure time. But in Aden, there was no enemy radar and no effective ground-to-air threat, which meant that flying low had little if any tactical advantage, except perhaps surprise. Indeed, flying low consumed more fuel and made our aircraft more, rather than less, vulnerable to rifle and small-arms fire. But such was the current wisdom of the day, and anyway it was definitely more fun!

It was my DFGA squadron, No 43 (Fighter) Squadron, along with its sister Hunter squadron, No 8 Squadron – both units of Khormaksar's Strike Wing – that would be called in by Army or Commando units when support was needed to defend themselves from ambush or to prosecute an attack. There were a dozen pilots on each squadron, and we would alternate our periods on duty: one squadron on duty from dawn to midday; the other from midday until nightfall, and vice versa. (Dawn and dusk in the tropics are always rather abrupt and always around 6 am and 6 pm respectively throughout the year.) This meant that each squadron would generally be stood down either for the morning or for the afternoon every day. And for us junior officers with little administration to attend to, this meant a trip to the beach club or some other recreational activity. To spend an afternoon lazing on the beach after returning from operational missions carried out just hours before might seem a bit incongruous, but it was a necessary relief. This split-day routine had been adopted to mitigate the cumulative fatigue of intense operations, heat, and humidity, not only for us pilots but also, perhaps more importantly, for the ground maintenance crew, who most definitely had a harder time of it than we did. Pilots would generally be programmed to fly two flights each in a half-day shift, each flight – from walk-out to walk-in – lasting about two hours, and each requiring perhaps an hour's pre-flight planning and half-an-hour's post-flight debrief. A typical 6-hour shift in those tropical conditions could knock the stuffing out of you – hence the occasional need for a bit of R&R.

Chapter 4

By way of armament, the Hunter carried four 30mm Aden cannon, each with 150 HE (high explosive) rounds and a firing rate of 1,200 rounds per minute, giving eight seconds of fire if all four guns were fired at once, or sixteen seconds if firing only two guns at a time. Hunters were also able to carry a range of external weapons on wing pylons including 1,000lb bombs, but these were never carried in my time in Aden. When called forward in support of ground forces, our standard weapon configuration was four Aden cannon and either eight or sixteen 60lb HE rockets.

Arriving within the operational area, our aircraft would generally come under the control of Forward Air Controllers (FAC) who would direct our fire to targets identified by the field commanders. These directions were not very sophisticated by modern-day standards against an enemy often dug in so well that they could hardly be seen from the ground let alone from the air. Even finding the location of the Army FAC could be difficult in such mountainous terrain, especially in those anti-cyclonic inversion conditions already described, when the haze could be as thick as smoke. Arriving at the FAC position, usually marked with a fluorescent-red fabric 'T' panel laid-out on the ground, its identification sometimes also aided by orange smoke, our aircraft would orbit overhead at a few thousand feet to receive instructions and survey the area. Unless the intended target was clear and obvious, the FAC would radio his targeting directions relative to this 'T', which might go something like: 'Target 2 o'clock my 'T', range 1,000-yards, 100 yards left of a bushy-topped tree' (if there were such a rare phenomenon in the area). One aircraft from our formation would then lay down a short burst of cannon fire on this position to confirm correct identification and act as a datum point. Depending upon where those first marker rounds were seen to fall, the FAC might then offer a correction, for example, 'From last fire: forward 500-yards, left 100-yards.'

The formation would now enter their attack dives, one after the other, firing on the position established in this way, with further corrections given if required. By the time all of our rounds and rockets had been expended, there would be so much smoke and dust hanging in the air that it was usually impossible to see from the air if the attack had had the desired effect. I'm afraid to admit that our part in such skirmishes always seemed rather clinical, somewhat remote from the immediate consequences of our bombardment on the ground. If we had carried out our job effectively, insurgent weaponry would have been neutralised or dispersed and our surface units thereby saved from injury or loss of life.

These events are uncomfortable to remember as I reflect upon that distant past with the clearing hindsight of my wise old-age, and they are not easy to talk about to an audience unaccustomed to such barbarity. In the light of more recent conflicts in the Middle East where British forces have been deployed, they are nevertheless relevant to recall. Servicemen and women are the instruments of political determinations not the instigators. Patriotic young service-men and women, bound together by a sense of duty to do what is asked of them for their country, respond loyally to the call-to-arms. It is they who put themselves in the line of fire and face the consequences of conflict; suffering the mental and physical wounds of it – even giving up their lives in the course of it. It is only later, after the heat of the battle has cooled and on sober reflection, that the survivors might question the sense of it all, and wonder if there might have been some better way. It seems, regrettably, to have been ever thus.

In my twenty-five-month tour of duty in Aden, I flew twenty-three offensive operations and 375 flying hours over the Protectorate. It is almost certain that small arms fire would have been directed at our low-flying Hunters, but this was never very effective against such fast-moving targets. However, since the treatment of captured British servicemen was known to be barbaric, we aircrew carried so-called 'ghooly chits' – yellow cards that promised (in Arabic) a reward if the carrier was returned with all his body parts intact. And some of us carried side-arms too (which were likely to be equally ineffective), in case we found ourselves downed in unfriendly territory. The only bullet to lodge in my wing, however, was one of my own 30mm cannon shells (fortunately an inert round used for gunnery practice), which ricocheted off a splash target being towed by HMS *Victorious*, then passing Aden en route to the Far East. Another occasion when I might have (but didn't) come under fire, was when I was once scrambled to intercept an Egyptian MiG 15 fighter attacking the political outpost in the Bayhan al Qisab area, about 125 miles north-east of Aden. This was the only occasion when my air-to-air combat training might have been put to use. I certainly felt the adrenalin coursing though my veins as I tightened my harness before take-off. Even as I neared the area after a high-speed dash at 25,000ft, I was receiving frantic radio calls from the controller of the Bayhan strip that the strafing attacks continued. But by the time I had arrived in the overhead with my guns armed and ready for combat, the MiG had departed. I therefore never got to try my skills in earnest (which may have been a very good thing), and the MiG had more likely fled the scene because he was running short of fuel rather than being scared off by my arrival.

Hunters (and sometimes Shackletons) would also be used for 'Flag Waving' flights over the Protectorate to show the British presence and to demonstrate the promised protective oversight to the sheikdoms of the new Federation. But much of the rest of our time was occupied with training (and more training) to develop or maintain our flying and fighting skills. These exercises mainly consisted of live air-to-ground gunnery or rocketry on the nearby weapons range, simulated air-to-air combat, and two-ship or four-ship hi-lo-hi simulated strikes on isolated targets hundreds of miles away up country. Our performance on the live weapon range would be assessed by the number of rounds hitting the targets, counted manually and sometimes embarrassingly displayed for all to see in the crew room. The rounds in each aircraft's magazine would be coated in paint of different identifying colours so that one pilot's hits could be distinguished from another's. Scores would be displayed in the pilots' crew room, so there was no hiding place and a pilot could end up with an expensive bar bill if he'd had a bad day, one or two of which, I have to admit, I had!

For simulated attacks – both air-to-air and air-to-ground – cine-cameras mounted on top of our gyro-gunsights recorded the accuracy of our aiming. You might find this difficult to believe but fighter pilots have been known to exaggerate, so these cameras were used to keep us honest! After each sortie, the PAI (pilot attack instructor) would invite us individually into his dark room to view our film and assess our aiming performance frame by frame. It felt a bit like being sat in the headmaster's study listening to a critical school report, because no performance could ever be quite good enough as far as he was concerned. If you were arrogant enough to think that you'd done well, it was his job to bring you back to earth with a bump and make you try harder. Grainy, black-and-white screen-shots from our firing passes were flashed up at the rate of one frame per-second, projected from an old 16mm projector that clattered through our cine-film in stop-frame mode. The PAI looked not only for the positioning of the aiming pipper at the point of firing, but also at the steadiness of the aim. To be awarded a 'hit', the pipper had to be steady on the target for several frames at least. In simulated combat, it also had to be held on your 'enemy's' forward fuselage, just behind his cockpit, for a 'kill' to be recorded. Less than steady or even slightly wide of the mark would be recorded as a 'miss'.

In hard-turning air-to-air combat situations, the aiming pipper of the Hunter's gyro-gunsight lagged below the bore-sight as if it were on elastic, which stretched with 'g' and the range from your opponent. This lag effectively represented how far ahead of the target you had to point your guns for your rounds to hit, making allowances for the muzzle velocity of the rounds, their time in transit, and the effects of gravity. The pipper also wandered from side to side if you were heavy-footed on the rudder pedals in your aiming, which in some rough combat situations was difficult to avoid. Tracking a manoeuvring target, therefore, felt a bit like trying to poke a cornered rat with a long length of bendy hose pipe; you were lucky if you achieved a hit rate more than thirty per cent. This is not as bad as it sounds, however. In real combat situations, just one hit in the right place on the target with a 30mm high-explosive round would probably have been be enough.

No 43 Squadron's PAI during my time on the squadron was Flying Officer Andy White, who would eventually become my best man and a good friend. I can still hear his scoring verdicts ringing out in the darkened room as the telling frames flashed up on the screen one by one – 'miss', 'miss', 'miss', 'hit' (if you were lucky) – intoned emphatically in a voice that would brook no dissent. His commentary would have reverberated through the thin walls of our prefabricated squadron hut and could be relied upon to raise some wry smiles among those gathered for the entertainment outside.

A Martin-Baker let-down

It was on one such training sortie that things went badly wrong for me. I was detailed to fly as number two of a battle pair on a low-level simulated strike training exercise. Our route was to take us from Khormaksar north-eastwards up the coast past Zinjibar and Shaqra, then inland up onto the Lawdar plain, where we were tasked with showing our presence to some of the walled cities up-country. As wingman in the formation, I carried out the take off in the starboard-echelon position, initially tucked in close to my leader's wing. Once airborne with our undercarriages retracting, I manoeuvred outwards into a defensive battle formation, placing my aircraft some 250 yards from, and 45° swept back from, my leader. This was to make the formation more manoeuvrable and also to give me, as number two, the freedom to look behind us into our vulnerable 6 o'clock sector from where any attacking aircraft might be expected to creep up upon us (although they never did). Entering a sharp left turn onto the required heading, we accelerated rapidly to 420 knots and settled at a height of 250ft above the ground. Breaking surf on long empty beaches stretched ahead, marking our route into the hazy distance towards Zinjibar.

. It was only a few minutes into our flight, already perhaps fifteen miles away from Khormaksar, that my engine RPM suddenly unwound without warning. It was one of those heart-stopping moments where one's first reaction is incredulity, to deny what is happening and to think that the instruments must be malfunctioning in some way. But there was no denying the absence of engine noise and the sudden loss of forward momentum. I put out a radio call to my leader, by now fast leaving me behind as my airspeed bled away. 'Gold 2, engine failure, turning back', I called as I pulled the nose up into a climbing turn, using what was left of my excess speed to gain precious height. The Hunter's best gliding speed is 210 knots (if I remember correctly), and I achieved this as the aircraft reached the apogee of its climb at about 2,500ft, just as I finished my turn. The craggy rim of Jabal Shamsan was now dead ahead in my windscreen, but I realised at once that the Khormaksar runway was too far away for me to reach without power.

My stricken Hunter was now descending fast, its Avon engine dormant. Even as I had been flying it on its turning course back towards base, I had also been trying to work out what had gone wrong – attempting several times to coax the engine back to life as my altimeter needle unwound like a clock on fast-rewind. But nothing I tried had made any difference – the engine seemed completely unresponsive. As the aircraft descended through 800ft, I knew that I would have to eject – any attempt to land a high-speed fighter on sand was not normally expected to lead to a happy outcome. I took a deep breath and made a final R/T call to my leader, whom I could see still orbiting above me, 'Gold 2. I can't relight – I'm baling out,' and I turned my aircraft away from the distant craggy peaks and levelled my wings. I had practised the ejection drill and the parachute landing technique many times, but only in the simulator and by jumping off boxes in the gym – now was my opportunity to try it for real. And if I wanted to walk away from this, I had no choice.

The Martin-Baker Mk2H ejection seat could be fired by either of two handles: top and bottom. The preferred handle was at the top of the seat, above the pilot's head. This double-looped black-and-yellow handle was combined with a face blind to shield the pilot's face from debris as the canopy was jettisoned, and to restrain his head from the buffeting it would receive once exposed to the 250mph airstream outside (it could be much faster, of course). The lower handle was located between the pilot's thighs; while offering no protection for the pilot's face and head, this handle was quicker and easier to reach when experiencing violent 'G' forces, for example, following a structural break-up or when departing controlled flight.

I chose the top handle, reached up with both hands, grabbed it and pulled it down over my face. There was a loud bang followed by the deafening roar of the slipstream as the canopy was torn away. An instant later, I was unceremoniously ejected from the aircraft at an acceleration rate of twenty-two times terrestrial gravity. For a fraction of a second, therefore, my body weighed twenty-two times its normal weight. It felt like a monumental kick up the backside, worse than anything I had experienced before, worse even than being set upon by the heavyweight pack of the Evesham Rugby Club. I heard myself grunt as the wind was ripped from my lungs, but a second later I became a mere projectile, still harnessed tightly to my seat, and following a ballistic trajectory like a shell fired from an artillery piece. I tentatively lifted my face blind. Between my feet, I could see my Hunter flying off, maintaining the steady course that I had trimmed it for, its cockpit now an empty hole. I could see the explosive rod that had so brutally propelled me upwards, sticking up like a finger pointing in the direction I had gone.

A moment later, after that briefest of brief respites, all hell seemed to break loose. With a pre-set delay, the automatic seat-separation sequence had been triggered; the main parachute was now being pulled from its housing by the stabilising drogue-chute. Suddenly, I was being yanked roughly from pillar to post as first the seat detached itself from me and fell away and then the parachute began to inflate. In a relative wind-speed three times that of a full gale, the canopy expanded rapidly to its full extent, its lines yanking, stretching, straining – throwing my arms and legs around like a rag doll. At some point, as my limbs flailed, I must have let go of the handle that had triggered the wild ride that had just saved my bacon, for I realised that I was no longer clutching it. I remember feeling an instant of disappointment to have lost this memento of the occasion. But a few seconds later, this became the least of my concerns, for I realised that the ground was rushing up to meet me at a frightening rate.

My parachute training had taught me to adopt a particular posture for landing – legs held firmly together, knees slightly bent, feet turned sideways to the direction of travel, hands holding the parachute risers, and arms and elbows up close around my ears to protect my head. The theory was that this posture would spread the impact with the ground along the side of my legs and around the back of my torso as I rolled over. We had practised this a hundred times on rubber mats in the gym. But this was not the kind of parachute that sport parachutists use, which might have allowed such finesse. This was an ejection-seat parachute, cut down to the minimum size to reduce weight; and the resulting descent rate was equivalent to free-falling from a 13ft wall. I landed like a sack of potatoes hurled from the back of a lorry! Despite my perfect posture, my legs simply buckled beneath me as I hit the ground so hard that it made my teeth judder. Tumbling forward, I fell spread-eagled, my arms outstretched, my bone-dome helmet grinding in the sand. I saw stars circling in a black sky, then felt the parachute settle over me gently like a shroud. I opened my eyes. Instead of blackness now, I was enveloped in a silky whiteness – I might have ascended into the clouds again, this time with my harp, I mused – but, moving my limbs tentatively one by one, I found myself entirely in full working order.

Clawing the parachute from my face, I gingerly got to my feet and checked my state. Nothing broken. I looked about me while unbuckling my harness and gathering up my parachute into a bundle. I could see the peaks of Shamsan in the distance. Swinging my gaze, I saw a column of black smoke rising from the burning wreckage of my aircraft. For a moment I worried that it might have caused casualties on the ground, but it was soon clear that my aircraft had somehow found a route through the forest of tall masts of the aerial farm a mile or so from where I now stood. I removed my flying helmet, noticing the deep gouge marks in the cheeky caricature of the 43-Squadron fighting cock painted on its bone-hard surface; I would need a new one, I thought idly. I felt battered and winded by my landing, but not hurt. The noise of a jet-engine above me caught my attention, and I looked up to see my leader's Hunter still circling – John Osbourne reassuringly keeping an eye on me – he would have radioed my position to base and so I knew that I would not have long to wait. I was alone on this empty stretch of flat, sandy scrubland, and it was unlikely that I would be in any danger from hostile locals so close to base. I spread out my parachute to make it easier to be seen from the air, and in doing so, a flash of something yellow caught my eye – something lying on the ground about a 100 yards away. I recognised it immediately as the inner-lining fabric of my ejection-seat face-blind and I ran over to retrieve it; I had got my souvenir after all!

The search and rescue Whirlwind was with me within minutes and it landed beside me with its rotors still turning, its piston engine sounding like a tractor's, running at a fast-idle RPM ready for an immediate take off. I bundled up my parachute and harness again and leapt into the aircraft's cavernous rear hold, aided by the crewman who strapped me in. The trip back to Khormaksar was quick, noisy and rough, but in less than ten minutes I was being deposited outside the medical centre to be greeted by the station medical officer.

'How d'you feel young man?' he shouted above the clatter of the Whirlwind's whirling rotor blades as the helicopter took off again.

'Fine,' I shouted back, perhaps a little too jauntily in my elation to be on the ground in one piece.

'How's your back feel?' he asked, as we walked into his surgery.

I tentatively twisted my torso from side to side and shrugged my shoulders. 'Perfect,' I said.

'OK, touch your toes.'

I complied readily, wanting to demonstrate my agility.

'Seems fine to me,' I said, straightening.

The SMO studied me carefully.

'Hmmm,' he muttered, not sounding quite so sure. 'Better get you down to the hospital for an X-ray, just in case.'

An open-topped Land Rover was duly summoned and I leapt into the passenger seat alongside an RAF corporal driver, who then raced me down town to the military hospital perched on the rocky slopes above Steamer Point. It was a rough four-mile ride on the Land Rover's bullet-hard suspension, and it was not long before I began to feel myself stiffening up. I was soon wincing with every swerve and jolt, bracing myself on the windscreen arch to take the weight off my back. By the time we arrived at the hospital's front entrance, I could hardly move. In fact, I needed to be lifted out of the vehicle bodily and carried in on a stretcher. Predictably enough, an X-ray quickly confirmed that, like most users of Hunter ejection seats, I had sustained a compressive fracture of L1, one of the lower vertebrae in the lumbar region of my spine, and would need to be immobilised while it healed. The delay between ejecting and my seizing up, was an example of the body's ability temporarily to mask pain by flooding the nervous system with adrenaline – undoubtedly intended by our maker to allow us time to flee from further harm following injury or predatory attack. In my case, it must have taken a good hour before my back injury made itself felt. Had I not been rescued, I would certainly not have been able to 'flee' for more than a few miles. So much for the prospect of successful escape and evasion following a bale-out using one of those early ejection seats!

34

CHAPTER 4

I spent the next three weeks in hospital, the first strapped to a hard board while my back settled, then two more with increasingly rigorous physiotherapy. My bed was placed out on the verandah with a glorious, elevated view of the harbour – a sparkling, colourful, busy scene filled with little boats and ancient dhows and cargo ships resting at anchor. The flight-path to Khormaksar's runway 08 lay across the bay; a plane-spotter would have had his hands full making notes of all the aircraft coming and going. The Hunter formations were the most numerous of all, as well as the most noisy, and I wished I could be up there with them again; but it felt good to be on the mend, even though I would be grounded for a while.

In the bed alongside mine was a Royal Marines lieutenant, an armoured-car troop commander – in for a circumcision of all things. The surgeon's sutures must have tickled his nerve-endings because, for several days after his operation, he had a permanent erection. He was given a pocket-sized ethyl-chloride freeze spray to shrink his prominence and save his embarrassment. When he joined us other invalids in our dressing gowns at the ward dining table for our meals, he would first have to duck behind a modesty-screen to administer a targeted blast of icy air; it allowed him to walk to his seat without looking like a bell tent! For the rest of us inmates on the verandah, it was the psst-psst-psst sound of that aerosol that we came to associate with mealtimes, and it always caused wry glances between us as we waited for him to appear. The lieutenant and I were discharged on the same day and we flew back to the UK together on a British United Britannia trooper. His wedding was planned to take place a few weeks later and I was invited to attend. He caught my eye with a knowing look as he mentioned his time in Aden in his wedding speech; I'm sure that we were both thinking of that aerosol!

I arrived back in Aden just before my 21st birthday, and so decided to throw a party to celebrate both events. No 43 (F) Squadron never needed much of an excuse for a party, but the timing of mine was helpful since we were mourning the recent loss of Ian Stevens, a fellow graduate of my Vampire course at Linton-on-Ouse, who had joined the squadron with me. He had unaccountably crashed into the sea just after taking off from one of our regular squadron detachments in the Oman. My party would thus also serve as his wake. For this I booked the Officers' Mess Jungle Bar, so called because it resembled one, its lush vines and palm trees flourishing on diligent watering in the tropical heat. Even on a Flying Officer's meagre salary, I was able to afford a consignment of crayfish tails flown in from Masirah, using the regular RAF Argosy shuttle-service that kept that remote RAF airbase supplied.

The crayfish curry buffet that evening was not only memorable for its incendiary qualities, it was also my first encounter with Caroline, an Aden

Airways stewardess on Vickers Viscounts and Douglas DC3s – and my future wife and partner in life. She wore a simple, sleeveless white dress cut just above the knee. Against her bronzed complexion, the whiteness of her dress material was dazzling. Her long dark hair, swept back and tied up in the form of a loose bun, was designed, I think, to be a perch for her forage-cap when on duty. She was a stunner. We exchanged no more than a few polite words of greeting before she and her partner disappeared into the throng, and I don't think I spoke to her again that evening. In fact, I would not do so again until some months later, following the return of the Squadron to Khormaksar after a deployment to Masirah in the Oman.

Chapter 5

Masirah Island lay about a thousand miles north-eastwards along the coast from Aden, and the Squadron deployed there from time to time for exercises over the Oman desert. These deployments also served as a change of scene from the Aden Protectorate territory with which we had become almost too familiar, although actually it was just a different kind of sand! This barren little island lay just off the south-east corner of the Arabian Peninsula at the entrance to the Straits of Hormuz. It was a stranded staging outpost, with a single runway and an assortment of corrugated aluminium buildings, which was staffed by a few lonely RAF personnel on a one-year unaccompanied tour of duty (some say as punishment for a misdemeanour of some sort), and roamed freely by feral goats. These animals got in the way from time to time when they strayed onto the runway, but occasionally captured, skinned and cooked, their meat made an outstanding curry! We spent our days flying around the rolling sand dunes of the Empty Quarter, and our evenings in the bar listening to Bert Kaempfert's Swingin' Safari over and over again, until it drove me quite mad.

Just prior to returning to Khormaksar, I'm afraid that I blotted my copybook performing an illegal beat up of the airfield, for which I was given a severe bollocking by our CO. As punishment, he scheduled me for three extra week-end duties as Station Orderly Officer, which effectively confined me to base for three consecutive Saturdays and Sundays, day and night, attending to any minor issues that might arise (the major issues would be taken care of by the more senior Station Duty Officer). Moreover, I would have to spend those weekends uncomfortably over-dressed in my number-one best khaki uniform with a Colt 45 pistol and holster strapped around my waist. My CO clearly meant me to learn a hard lesson about flying discipline from that, I think, and I certainly did.

It was during the evening of one of those punitive Saturdays, as I stood in the reception lobby of the Officers' Mess dressed in my uniform, looking military and purposeful, that I was to bump into Caroline again (or more accurately, she was to bump into me). I think that she must have been returning to the Jungle Bar after a trip to the Ladies Room when she came upon me. I acknowledged her with a friendly, non-committal smile as she passed by, not expecting any more than a friendly glance in return. But instead she stopped and we talked for a while …

A week or so later, as I was driving my MG Midget back to Khomaksar for the afternoon shift after a morning at the beach club, I caught sight of her in my rear-view mirror, coming up fast behind me in her red open-topped Alfa Romeo Spider. As my hood was down too, I realised that she would have recognised me easily even from behind, and I slowed to let her catch up. In my mirror, I saw her give me a little wave as she closed the gap between us. A dinner-dance was scheduled in the Officers' Mess that coming weekend, and I wondered if I should risk asking her to accompany me. Without dithering any further, I made my mind up, and seizing the moment, I indicated that we should both pull off the road. She drew her Alfa up alongside me on the hard sand of the verge, and without leaving our vehicles, I made my proposal. I had steeled myself for a polite refusal, but she accepted with one of those enigmatic smiles that I would come to know so well. We were engaged within a month …

Six months later, I was back in Masirah again for another ten-day spell of flying over the Empty Quarter of the Oman with excursions to other RAF outposts in Sharja and Bahrain. It was during this detachment that I was detailed to ferry one of the Squadron's Hunters (XV609) back to the UK for refurbishment. My first port of call was Bahrain where I would meet up with a couple of pilots of No 208 Squadron, then resident there, who were to accompany me flying two other care-worn aircraft of the type. It was to be a three-day journey back to the UK, taking us via Tehran, Akrotiri (Cyprus), and Malta, to the maintenance unit at RAF St Athan, where the refurbishment of all three Hunters would take place. We took off in the early twilight two days later, and by the time we reached our cruising altitude of 40,000ft, the sun's incandescent rim was glinting on the horizon. It soon rose higher and bathed us in its orange light as we flew northwards in wide formation across the mountains and deserts of Persia (as we called it then). On the ground, dawn was still to break, yet our condensation trails streaked the clear blue sky behind us with three iridescent amber lines that signalled the coming of the new day to anyone below who might lift their gaze. Soon, as the sun continued its ascent, every earthly detail threw long shadows across an arid landscape of rocky ridges and moon-like plains. Here and there, a few small settlements lay in what seemed a sparsely populated land – towns and villages linked by dusty roads through mountain passes, and edged by irrigated stripes of variegated green. The dry air was as transparent as crystal, and Tehran was already visible with fifty miles still to go. Behind the city lay the Elburz mountain range that stretched East-West from horizon to horizon, and beyond its snow-capped peaks, lay dark, densely forested slopes that ran all the way down to the Caspian Sea. We had a grandstand view of the country in its best light that morning, and that picture of it has stuck in my mind all these years.

This first leg of our journey took about ninety minutes from our take-off in Bahrain to our landing in Tehran. Marshalled to park our Hunters on the military hard-standing a discreet distance away from the commercial parking lot, we were picked up by a chauffeur-driven limousine and delivered to the airport terminal. I remember passing through the fountains and ornamental pools of the airport's entrance with my eyes on stalks, never having seen such oriental opulence before. We felt rather out of place in our military flying overalls as we were ushered through the airport's marbled halls to join the airline crews wearing gold-ringed jackets and fresh white shirts. But we soon got over any feeling of inferiority – after all, our fast jets were much more exciting beasts than theirs to fly, and the envy was clear upon their faces. Our ground-handling agent, a local man assigned to us by the airport authority, treated us like royalty from the start, not only organising our refuelling and assisting us with our onward flight-plan, but also by laying on a lavish breakfast at a private table in the VIP suite. It seemed a far cry from the sink-full of dirty coffee mugs that we had left behind in 208 Squadron's crew-room from which we had departed less than three hours before. Unfortunately, this cosseting was all over far too quickly – by mid-morning we were lining up again for take-off on Tehran's runway 29 with 13,000ft of black tarmac stretching out before us.

Taking off in close formation as a threesome, we climbed to 30,000ft and settled at a cruise speed of Mach 0.85, flying in a wide 'V' pattern about a thousand yards apart. Our leader, Flight Lieutenant 'Hoppy' Granville-White, flying at the apex of the 'V', did most of the navigation and handled all the radio calls, so we two wingmen had a pretty easy time of it. We were about half way into our 2½-hour flight to Akrotiri, when I noticed a stream of white vapour issuing from my fellow wingman's aircraft. At first, I thought it was a condensation trail, but seeing none behind my own or the leader's aircraft, I knew that it could only be a fuel leak. These were old Hunters that had been subject to pretty punishing conditions in the abrasive air of Arabia, and it was not surprising that their systems were beginning to fail. I slid my aircraft across the formation to take a closer look. Fuel streamed from the aircraft's starboard wing-root – a vapour trail issuing like a thin version of the smoke used by aerobatic teams for painting figures in the sky.

Fifteen minutes dragged by as we monitored the leaking aircraft's fuel usage and compared it with our own. Our route had taken us north across Turkey to avoid overflying Iraq and Syria, and this long way round meant that fuel was tight. If the leak turned out to be major, we would have to divert, and landing at some remote Turkish airstrip for a repair job was not something we were very confident about. Thankfully, as the minutes ticked by, it became clear that the leak was not critical, but it was another anxious hour before we were sure we would make it all the way.

We hoped that a quick patch-up job at Akrotiri might get the aircraft home where it could be properly attended to, but an engineering inspection revealed that the leak was too serious for that. The fuel tank would need replacement before the aircraft would be cleared to fly again. This unwelcome news triggered a flurry of telexes between ourselves and home base – the main question being: should the two of us with serviceable aircraft continue on, or should we all remain in Akrotiri until the incontinent Hunter was repaired? For some reason, the distant powers-that-be decided that it should be I who would remain behind, while my two colleagues would continue on. To say the least, I was a little miffed when I was told this, and even more miffed as I watched the pair take off the following morning without me, taking my perfectly serviceable aircraft with them. I began to think it was a 208-Squadron stitch-up! Moreover, it was soon revealed that to replace the faulty fuel tank, the entire wing of XF431 would have to come off, which would require specialist lifting equipment and an engineering team to be flown in from Bahrain. The forecast time for work completion was two weeks.

It didn't take long for me to realise, however, that if you are going to get stuck anywhere in the world with an unserviceable aircraft, getting stuck on a Mediterranean island in springtime is not the worst of all situations. So I made the best of it, spending a couple of days touring the island in a hire car and the rest of the time with the RAF Akrotiri sub-aqua club learning to dive, something that I had always wanted to do.

Once cleared to go again, I flew my Hunter to RAF Luqa in Malta, accompanied by a No 39 Squadron PR-9 Canberra to shepherd me. This was a precaution insisted upon by Middle East Command to safeguard its assets. In my Hunter, I only had a map, compass, and stopwatch to navigate by; and except for Crete about half-way along my route, my aviation chart showed nothing but pale-blue, which did not offer a lot of opportunity for pilot-navigation! Without the Canberra's long-range radio and navigation aids, finding a small island like Malta in the middle of all that water would have felt just a little bit precarious with limited fuel. Anyway I was glad of its company. Flying for so long over so much sea with only one engine was tempting providence, and it was comforting to know that my position would be reported should another engine failure force me into an inflatable rubber dinghy.

In the end, the trip turned out to be non-eventful, taking two hours and twenty minutes from take-off to landing. Except that around half-way along my route, a warning light in my cockpit told me that one of my undercarriage legs had become unlocked. Most pilots will admit that any in-flight warning causes a momentary increase in heart-rate. Over an apparently infinite sea, I can tell you that it roughly doubles. Fortunately the Canberra pilot was on hand to look me over and confirm at least that the door appeared in place, but after landing, the engineers in Luqa grounded me while they investigated the fault. This was to be another week off on a Mediterranean island with time to kill! This trip was turning out to be a bit of an adventure.

The next leg, from Malta directly into RAF St Athan, was the longest of the whole journey and it didn't start well. Not long into the climb, the fault occurred again. This time I was pretty sure it must be a microswitch fault rather than anything more serious, and so I chose to ignore it and skip my intermediate fuel stop in case the fault grounded me again. I reckoned that I could make it, and the best place for my clapped-out Hunter was in the maintenance hangar at St Athan where this fault could be properly sorted out. But it was a long trip, and my fuel was tight all the way. It was great relief, therefore, when I was able to close the throttle over Normandy to begin my descent. From 30,000ft, the Hunter could 'glide' a good hundred miles at idle power, and my throttle stayed firmly closed until I levelled out at 2,000ft over a cloudy Bristol Channel. It was at this point that I switched radio frequency to St Athan's control tower, only to be told that their direction-finding equipment was unserviceable. The visibility was poor and the coast was partly obscured by patchy cloud and rain at the time, so it was not obvious which way to turn. And with only minutes of fuel remaining, I could not afford to get it wrong. What else could I do but trust my homing instincts; I turned 90° to starboard and held my breath. If I had not chosen correctly, I might face a parachute descent into the muddy waters of the Bristol Channel, now sliding ominously below me. I spent the next half-minute or so (though it seemed much longer) anxiously peering into the grey murk of a rainy South Wales day, until the approach lights of the runway eventually appeared in my windscreen. I was quite relieved to hear the squeal of rubber on the tarmac as I made my touch down. My flight had lasted two hours and fifty-five minutes, an awfully long time to be sitting on the unyielding seat-pad of the Hunter's ejection seat – and my bottom ached!

Two days later, I was on my way back to Aden in a British United Britannia trooper. My squadron commander clearly thought that I had been away long enough!

Chapter 6

Back once again in the sandy streets of Aden, I found myself drawn in to Caroline's circle of Aden Airways friends and colleagues, and my social life soon began to change – one distinctly more varied than my former RAF-bachelor existence had been. My social orbit until then had been entirely centred on the camaraderie of fellow squadron pilots, a very masculine, fighter-pilot ethos that demanded a certain amount of swagger, a lot of noise in the bar, and the consumption of quite a large quantity of beer. It might be some excuse to blame this on the almost entirely male community of RAF service-life in those days, but the number of unattached females in the Colony was so few that there was little opportunity to expand one's social orbit. As Caroline's fiancé, however, my social life now included new civilian friends who we would meet at the civilian beach club at Gold Mohur and in Aden Airways gatherings. One new friend, Jonny Rose, a former Hunter pilot himself who had left the RAF after a short-service commission, was now an Aden Airways DC3 captain, and I somehow managed to get him airborne in a two-seat Hunter T7 just for old-time's sake. In return, he put me in the co-pilot seat of his DC3.

Aden Airways operated Vickers Viscounts and Douglas Dakotas (DC3s) all around Arabia, East Africa, the Sudan, Ethiopia, and Egypt. The Viscount fleet, on which Caroline was a senior stewardess, flew on a weekly schedule that included Cairo, Jeddah, Hargeisha, Khartoum, Nairobi, and Bahrain. The DC3 fleet flew mainly into the up-country desert strips of the Hadramaut and across to Djibouti and the red-dirt runways of Somalia.

As already mentioned, the two Hunter squadrons of Strike Wing alternated their periods of duty between morning and afternoon, which meant that we had a half-day free every day to pursue other interests. Week-ends and leave added more free-time when we could do much as we liked. Being keen for any flying experience available, I pestered my new friend for more flights in his DC3s, and soon found myself such a regular that I was included in the Aden Airways duty roster as his second pilot. How this was possible or even legal, with me an RAF pilot in the service of Her Majesty and with no commercial flying licence, I will never understand. Whatever the airline's logic, I soon found myself performing landings on sand strips in the most remarkable places of the Hadramaut and in Somalia, passing myself off as a civilian co-pilot. Except for the blue stripes of my RAF epaulettes (rather than the gold stripes worn by commercial pilots), my uniform was identical to the Aden Airways uniform (khaki shirt, shorts, knee-length socks and desert boots), and so I doubt that any of our Arab passengers or the ground staff at these desert outposts noticed. This was probably just as well, because there would undoubtedly have been some local people in those up-country locations who might have been quite hostile to the presence of a military pilot in their midst – especially one who, not far away in the Radfan mountains, was often engaged in operations against their rebel countrymen. And I doubt that the yellow 'ghooly chits' that I anyway neglected to carry on those occasions would have saved my body parts if I had been rumbled. But such was my naivety that this risk never occurred to me, and I took every opportunity to fly with Aden Airways without a second thought.

I finally walked Caroline down the aisle of Holy Trinity Church, Brompton, in December, 1965. I had arrived back in the UK from Aden only nine days beforehand, having been scheduled for one of those periodic Hunter simulator emergency training courses at RAF Chivenor that we Hunter pilots had to suffer from time to time. With just a day or two free before its start, I managed to find a church and a willing vicar available to conduct the ceremony under a special licence from the Archbishop of Canterbury (which, amazingly, arrived in time). With all this hastily organised, I rushed off down to Devon to start my course. Caroline arrived in the UK with just three days to make arrangements for the service and reception while I, meanwhile, was being tested to destruction in the Hunter simulator, coping (and sometimes not coping) with dire and unlikely simulated emergencies. We returned to Aden after Christmas for the last four months or so of our tours, living in a rented flat in the noisy back-streets of Steamer Point, which we were not sorry to leave when the time came in May 1966.

The days of Aden's British administration and the Federation of South Arabia were numbered too; indeed, their tenures would soon come to an abrupt and bloody end. In Yemen, the royal family and their Saudi backers would be replaced by a new Communist-backed regime. The Egyptians, defeated in the Six Day war with Israel in 1967, would withdraw from

Yemen completely soon-after, and President Nasser would abandon his Arab nationalist aims. After the British withdrawal, the FSA would also crumble, and by the end of 1967 nothing of the previous administration was left. British forces would have endured the hardest anti-terrorist campaigning since Malaya, and the indignity of their departure would rank with its retreat from Palestine. There would be no grand ceremonies of independence in Aden. Instead, the British High Commissioner would have to be airlifted to a Royal Navy aircraft carrier standing well offshore. He left behind the People's Democratic Republic of South Yemen to take control of a city racked by fear and poverty, a city that came quickly to resemble a ghost town. 'Gone Away – No Milk, No Papers,' was the message left by a soldier on the wall of Aden's vacated prison. It was a poignant epitaph for a hundred years of British rule. And sadly, as we know for that unhappy country, worse was yet to come …

At its peak, 30,000 British personnel served in Aden alongside 15,000 Adeni troops of the Regular Federal Army. Losses of British service personnel during the campaign amounted to ninety killed and over 500 wounded. Losses of the Federal Army amounted to seventeen killed and fifty-eight wounded.

Chapter 7

When the time came for us to leave Aden in the spring of 1966, I went on ahead with a posting to the Central Flying School at RAF Little Rissington with clear instructions from Caroline to look for a 'cottage in the country'. This was the mission that I at once set myself upon in the weeks before my course began, scouring the countryside around the Cotswold airbase for a suitable property. A six-month summer letting of a 'pretty cottage with roses round the door' (the essence of my instructions) was hard to find, especially at short notice. But luckily I stumbled upon Far End Cottage in Kingham, which, from the outside at least, fitted Caroline's specification exactly.

My course started in early June, and I attended the informal meet-and-greet happy-hour in the bar of the officers' mess on the first evening. I had joined No 234 Flying Instructors Course, and with my Hunter background, I quite expected to train as a Jet Provost or Gnat instructor, which would then have led to a posting to one of the RAF's half-dozen basic or advanced flying training schools. But while mixing with my fellow students in the bar that evening, I fell into conversation with one of the staff instructors on the 'Waterfront', John Merry, who was touting for volunteers to join the Chipmunk instructors' course instead. His pitch was that upon graduation from CFS, Chipmunk instructors would be allocated posts at one of the UK's university air squadrons, of which there must then have been close to twenty covering the university cities of the four nations. University Air Squadrons' raison d'etre was mainly to promote what the Air Ministry termed 'air mindedness' among those who were likely to rise to influential positions in government and industry. A programme worth spending money on, the Air Ministry reckoned, to create a core of prominent people in the country sympathetic to, and well informed about, the RAF – people who by dint of UAS experience would be likely to contribute positively to any debate on the Service's future.

The instructional duties on these university squadrons, John told me, were different from those of main-stream military flying schools. UAS instructors, he said, must be ready to become part of university life as pseudo-tutors to their undergraduate students, and would not only teach in the air, but also teach aeronautical and aero-engine subjects at weekly tutorials. Moreover, since their students would not, in general, be as disciplined as military cadets, nor necessarily interested in service or aviation careers, UAS instructors could expect debate and challenge from their students rather than deference. John added (tongue in cheek, as his often was) that UAS instructors therefore had to be special kinds of people: inspiring, adaptable, and open-minded. His recruitment pitch was, of course, largely salesmanship, but I found myself attracted to this different, semi-civilian way of life in a university city. The following day, I signalled my readiness to transfer to the Chipmunk course; and as things were to turn out, I think that this was one of best decisions I ever made. John Merry was my tutor throughout the course until I graduated six months later with the De Havilland Prize for display aerobatics and a posting to Oxford University Air Squadron. As I look back on my time under John's guiding hand, I see him now as one of only a handful of people in my RAF career to have truly inspired me.

'I walk on air and contemplate the sun' – translated from the Greek inscription on the OUAS crest.

I arrived at Oxford University Air Squadron on a sunny November morning in 1966, aged not much older than the undergraduates that I would soon be teaching. OUAS, like all the UASs, was part of the Royal Air Force Volunteer Reserve, and its eighty or so undergraduate students had joined either as officer cadets or as acting pilot officers. Of the thousand or so applications that the Squadron received each year following the university's Freshers' Fair, it might short-list a hundred for interview, from which only twenty or thirty would be selected. Our students were attested into the Service (RAFVR), wore uniform on duty, and agreed to spend at least half a day a week at the airfield for flying or ground training. In addition, they undertook to attend weekly lectures on Wednesday evenings and an annual two-week summer camp. This was not much to demand for such a privilege, was it?

I joined a staff of five full-time RAF flying instructors, each of us allocated a dozen or so students to tutor and teach. Our CO, a wing commander, seemed to spend much of his time on administration and so had no permanent students, but instead flew with a selected few from time to time to check progress. The Squadron inventory and personnel strength included six Chipmunks, a minibus with civilian driver (Mr Dingle), and a dozen RAF ground-servicing technicians headed by a senior NCO. As well as its hangar and flight-training facility on the old WWI grass airfield at RAF Bicester, OUAS also occupied a spacious barn-like building in Manor Road, Oxford, clad incongruously in corrugated-iron among the classic dreaming spires. This building, a short walk from the city's centre, served as its town headquarters, social club, and administrative centre. It had a lecture hall, a bar and lounge, and a dining hall sufficient for large and formal gatherings.

Students could remain members of the Squadron for the three or four years of their undergraduate courses and most would be keen to learn to fly. Some would lose interest and some would fail to make the grade, but if they stayed the course, students could expect about 120 flying hours across a full flying syllabus similar to that enjoyed by their military counterparts in RAF basic flying-training schools. It was a remarkably good deal for those lucky enough to be selected: not only were they taught to fly cost-free, but they were also paid a per-diem rate when on duty. Those who had signed up to join the service after graduation, moreover, even received a salary, and it was these young men who would arrive at Bicester driving their own vehicles, while most others would have to suffer the journey in Mr Dingle's minibus.

We instructors would receive a fresh batch of students each morning and afternoon, of which two or three would be mine. Being time-tabled usually for three or four dual instructional flights a day with corresponding briefing and debriefing sessions, I was therefore quite busy. And when not actually in the air, I would be briefing and authorising student solo flying, or writing progress reports in their individual training logs, or watching the weather, or attending to my secondary duty of maintaining the Squadron statistical records (which I found a real chore). Wednesday evenings were reserved for ground-school classes, when I taught (or at least attempted to teach) technical subjects to very bright science and engineering undergraduates who sometimes technically knew more than I did. And every month there would be the pleasure of a formal black-tie dinner night in our Manor Road Mess, often with a guest speaker. We also took our students away on detachment for intensive bouts of flying during the summer vacation, exchanging our airfield and offices with those of another UAS.

It was a truly satisfying and rewarding tour, and for the first time in my albeit brief RAF career so far, I felt as if I was doing a grown-up job with a constructive output (unlike my time in Aden, which had fallen somewhat short of this). Moreover, being part of a mixed Oxfordshire community with connections to a university city was a highly stimulating experience.

'Imprimis Praecepta' – Our teaching is for ever. Motto of the Royal Air Force Central Flying School

My time in Oxford, however, would be relatively short. In April 1968, the death of one of the CFS staff instructors in a flying accident created a vacancy on the 'Waterfront' at Little Rissington that I was called to fill at short notice.

Until this posting arrived quite out-of-the-blue, I was already beginning to think about applying for test pilot training, which was an eleven-month course at the Empire Test Pilot School, then based at Farnborough. The seed of the idea that had been planted during my short secondment to RRE Pershore some years before had begun to take root. Indeed, I had artfully led a visit to ETPS with a small party of my Oxford students (to satisfy my own interest as much as theirs), and had been inspired by what I saw. But I soon found my new line of work as a CFS instructor so addictively stimulating that I decided to hold off my application for a while. I was also far from certain that I would get through the ETPS selection process, which involved two days of examinations and interviews and an entry standard of good A-levels in mathematics and a relevant science. Since I had not taken any 'A' levels at all, I knew that it would take me some time to prepare. I decided that my application would therefore have to wait until I was ready, and in the meanwhile, I threw myself into my new role.

A typical day's work for me on the 'Waterfront' at Little Rissington would include four instructional flights, each of about seventy-five minutes duration, generally split into two parts. In the first part, the student instructor would try out his instructional delivery on me in a simulated flying lesson that I had demonstrated to him during our previous flight. While he 'pattered' through his lesson, I did my best to act and fly like 'Bloggs', a basic flying student of average ability (and sometimes a lot worse). This part was what we called the 'give-back', which he would have practised with a fellow student on a 'mutual' flight previously, having spent several homework hours preparing for it and rehearsing his patter. In the second half of the flight, we would change roles – I would then demonstrate the next lesson in the flying syllabus (the 'give') while he would act as Bloggs. This 'give'-'mutual'-'give-back' routine would continue for the twenty-two flying lessons of the basic instructional syllabus, starting with the first 'effects of controls' exercise and progressing step-by-step to the more advanced formation, aerobatic, and instrument flying lessons. Each of these exercises would require detailed briefings and debriefings, again with role-play as student or instructor. My students and I therefore spent a lot of time together, either in the air or on the ground. I found the work thoroughly enjoyable and rewarding. While this was a demanding course for our students, all themselves highly experienced military pilots, there was also a lot of laughter in the crew-room and we did not to take ourselves too seriously.

At any one time of year, there would be two six-month instructor courses running in parallel at CFS, one course three months ahead of the next. Student instructors on these two courses would spend half their time each day on academic studies and half on airborne tuition, the two courses alternating. I would therefore teach two students from each course in each half of the day, and these airborne exercises would be followed by the usual write-ups in their personal training record. If I got a break for lunch at all, it would be short – a hasty sandwich or a sprint up to the 'Greasy Spoon' flight-line cafe for a burger and chips. My time was pretty well fully accounted for – and I loved every minute of it.

The following year, pressed by my friend, neighbour, and flight commander, Peter Blake, with whom I car-shared each day to and from the airbase from our adjacent homes in Shipton-under-Wychwood, I successfully applied for transfer to a permanent commission on the RAF's general list. Until then, such a career progression had not been remotely in my game plan, for in Aden I had been quite taken with the romance of dirt-strip flying in the DC3 and had considered resigning my commission and joining the airlines instead. Now I was glad I hadn't. My role as a flying instructor, particularly one on the staff of CFS, had turned things around for me. And having dived headlong into the main stream of the RAF's general list almost by accident, I was struck by a sudden if somewhat late onset of professional ambition.

As 1969 dawned, I was invited to join the Chipmunk aerobatic display team, 'The Skylarks', led brilliantly by Barry Nelson and later Steve Holding; and I spent many glorious weekends of that summer performing at various air shows in the UK and in Germany. Unlike the Red Arrows, whose immaculate display of nine high-speed Folland Gnats occupied a lot of airspace with their fast and slick manoeuvring, our four piston-engine Chipmunks were always very up-close and personal to the crowd.

CHAPTER 7

Instead of the roar of jet engines from the premier team, our Chipmunk's Gypsy Major engines back-fired, coughed, and spluttered every time our aircraft were inverted (which was quite a lot of the time). In a series of coordinated and dynamic aerobatic manoeuvres – loops and rolls and flicks and turns – flying as two close-formation pairs, we repeatedly flashed past each other at close quarters at the focal point of our display, creating a dizzying spectacle of movement and percussive sound, which must have sounded like a WWI dogfight.

We named our finale the 'Gruntwördle' (the word invented to sound appropriate for our German audiences), which commenced with our four aircraft diving at each other from opposite ends of the airfield, and then pulling up into a line-abreast vertical climb in the centre of the display area. Hanging in the air as our airspeed ran out at the top of this skyward swoop, the outside pair then performed opposite stall-turns, while the inside pair flicked left and right into the first turn of an incipient spin. It was spectacular for the crowd regardless of how well we performed it – which was just as well, since it was a very tricky manoeuvre to coordinate and was sometimes more chaotic than we intended. Nevertheless, we aimed to come out of our various manoeuvres crisply and simultaneously, all pointing vertically downwards, before diving like the guy ropes of a maypole to the four corners of the horizon. We usually got a lot of applause (and sometimes laughter), but we never cut quite the same dash as the Red Arrows with their nine sleek jets and their tailored scarlet flying suits. And while we might have drawn gasps from the crowd with our noisy cavorting around the skies with such apparent abandon, it was they who drew the most requests for autographs.

While all this was going on, I was setting myself some further professional targets that involved more late-night study. I have to admit that this target-setting was now becoming a bit of an obsession. I had already re-categorised to A2 at OUAS, and now, coached and encouraged by another friend and colleague on the 'Waterfront', Dick Snell, I became an A1. This was the top qualification in the military flying-instructional field, and it would be false of me to deny that gaining this hard-won accolade was immensely satisfying, especially since I had had such a stuttering start to academic study in my school days. In teaching, I seemed at last to have put those early scholastic failures behind me and had found a niche. It was one in which I would happily have stayed for a while had not the offer of another line of work arisen that was so enticing that it would have been folly to refuse.

'Learn to Test, test to learn' – the motto of the Empire Test Pilots' School.

I drove down to Boscombe Down in mid-July 1969 to sit my ETPS entrance exams and be interviewed by a panel of distinguished test pilots, including Captain Pat Chilton RN, the School's commandant and chair of the panel, and Squadron Leader Robby Robinson, the senior fixed-wing tutor at the time. I think that I must have done quite well generally in the exams, but I was stumped by one of the lengthy calculus questions in one of the maths papers. Rather than give up, however, I worked up a solution by going back to first principles, developing my mathematics on the page. I had been quite good at mathematics at school, but there had not been much call for it since, and I would have been a bit rusty had I not thoroughly prepared. This I had achieved simply by ploughing through an A-level mathematics text book page-by-page in the months preceding these tests. It seems that my approach to the calculus question in particular must have given the School's Head of Academic Studies the impression (falsely) that I was a strong mathematician, because I was earmarked for the annual exchange place at the US Naval Test Pilot School. Later that week my selection was confirmed by letter, and this sent Caroline and me scurrying for the atlas. During my interviews, I had blandly indicated an enthusiasm for the US-exchange opportunity if offered, but I had no idea until then where in the USA the school was located.

Chapter 8

After some helpful jet refresher training on Jet Provost and Gnat in my final months at CFS, I left Little Rissington at the end of 1969. In the three-and-a-half years I had spent in the flying-instructional world – at OUAS and at CFS – I had amassed 1,400 flying hours on Chipmunks! I had often told my students that if they could fly a Chipmunk well and perform three-point landings consistently on the spot in all conditions, they would have developed the basic flying skills to fly anything. Over the next twenty years of my life as a test pilot I would be doing just that, and thereby proving my own assertion.

My family and I flew to Washington in January 1970. Our first-born, Sonja, was then just 3 years old and Andrew, our son, just 21 months. The day following our trans-Atlantic flight, we attended the customary British Embassy briefings that were given to exchange officers and their wives arriving in a foreign land – which included the bit about being 'ambassadors for your country' and not annoying the locals too much. The currency exchange rate was then US$2.40 to the pound, making a US cent exactly equal to an old British penny (there were 240 pennies in a pound for those who've never heard of them), and I made arrangements for my pay to be transferred to a US bank. My monthly net pay then amounted to about £200 plus some allowances for overseas living, amounting to about US$700 in all, out of which we would have to pay all our costs, including private accommodation off base. It was clear from the beginning that we were going to have to be quite frugal throughout our stay.

After a rather dismal weekend in downtown Washington's Hotel Windsor, which was as featureless and uninspiring as the soup that bore its name, a Navy driver arrived in a government car to pick us up and drive us to the base. Naval Air Station (NAS) Patuxent River, Maryland, was situated some seventy miles south of the capital on a narrow neck of land bordered by the Patuxent River and Chesapeake Bay on its Eastern shores, and the River

Potomac on its West. Lexington Park was the satellite town that had grown up on the Naval airbase perimeter to serve its personnel, a typical American settlement of stores, bars, and franchised fast-food outlets.

The US Naval Test Pilot School at the Naval Air Test Center (I had better use the US spelling) at NAS Patuxent River provides post-graduate courses for pilots, NFOs (naval flight officers) and engineers of the US Navy and the US Marines to train as test pilots, flight test engineers, and programme managers for their respective aircraft and weapon-systems procurement programmes. The US Air Force trains its test pilots and flight test engineers at its own test pilot school at Edwards Air Force base in California, while US Army aviators selected for this level of training are sent to whichever of the schools will best suit their likely employment thereafter. Both the US Navy and US Air Force Test Pilot Schools have exchange arrangements with other national test pilot schools around the world, including the UK's ETPS at Boscombe Down and the French École du Personnel Navigant d'Essais et de Réception (EPNER) in Istres, with whom students (like me) are exchanged from time to time.

When I arrived at NAS Patuxent River in January 1970, I was the only foreign officer in the cohort of nineteen students joining Class 56, and one of only two British officers on the base at that time. Lieutenant Commander Ian Normand, the Royal Navy exchange tutor and Head of Flight Test at the School was the other. Many of my fellow students had seen recent operational service in Vietnam, and although none of them knew much about the UK's Aden campaign in which I had played the small part already described, I could at least swap war stories with my new colleagues and so gained some standing.

The US Navy's offensive air power was then (and remains) formidable. In 1970, the USN had 8,646 aircraft on its inventory, over 3,000 of which were classified as combat types. These were operated from fifteen aircraft carriers and a score of Naval air stations in the USA and abroad. The total number of aviation-related military personnel employed by the USN in 1970 was over 140,000, 14,600 of whom were pilots on active duty. I had therefore joined a flying service comparable with the entire Royal Air Force, yet this branch of the US Navy formed only a quarter the total US Navy strength of 672,000 officers and enlisted personnel at that time. In addition, the US Marine Corps, effectively an integral and complementary arm of American ship-borne air power, contributed a further 4,900 pilots and 62,000 aviation-related personnel and operated its own inventory of operational aircraft from fighters to troop transports.

CHAPTER 8

Evidentia per Intuitionem – Class 56 USNTPS Motto

My fellow test pilot and flight-test engineering students on Class 56 were military aircrew drawn from the USN, USMC, and US Army. Two civilian US Defence Department (DoD) engineers already employed at the NATC joined us to train as flight-test programme managers. We were grouped into four syndicates, three fixed-wing and one rotary wing. My syndicate, led by Clint Smith, an F4 Phantom pilot, included Walt Lawrence, an A4 Skyhawk pilot, Jim MacDonald, an A6 Intruder NFO (navigation and weapon systems), Dick Gallant, a DoD civilian engineer, and me. It was traditional that each Class should have its own badge and motto, and after some debate and banter in the bar on a TGIF gathering a few weeks into our course, we settled on ours. This was a depiction of the cartoon character, Dick Dastardly (from *'Dastardly and Muttley in Their Flying Machines'*), and the Latin motto 'Evidentia per Intuitionem' – intuitively evident – a phrase used too often by our principal academics lecturer, Tom Moore, whose mathematical formulae sadly never were. USNTPS courses ran for eight solid months with only one week off in the middle of the course that served as a recreational break.

The first two weeks of the course were spent in full-time study, with refresher classes in aerodynamics, thermodynamics, mathematics, report writing, data reduction techniques, and analysis methodology. After that, our days were split into two, much like at CFS, with two overlapping course cohorts running in parallel, one four months ahead of the other. Also like at CFS, half our days were spent in ground school and half spent flying, analysing our data, and writing reports. The practical flight testing phase was accompanied by briefings and demonstration sorties with our tutors, usually in the T1 Seastar (a navalised version of the well-known T-33 Shooting Star advanced jet trainer powered by an uprated 6,100lb thrust Alison J33 turbojet) or the T-28 Trojan (an advanced trainer and counter-insurgency aircraft powered by a Wright Cyclone radial engine rated at 1,425hp – more than eight times the power of my beloved Chipmunk!). With no trainer versions of the School's more operational types (and no simulators either), these introductions were all we got before putting our newly learned test techniques into practice solo on the single-seat/single-pilot Skyhawk, Crusader, and Phantom. Even though these aircraft were already in service, they were entirely new to me, and so my first flights were as close as a trainee test pilot could come to a first flight on a new prototype. I certainly felt as apprehensive!

The first four months of the course focused on the measurement and assessment of aircraft performance; the final four months on aircraft handling and flying qualities, including stalling, spinning, and other departures from controlled flight. All of our results were required to be presented in a standardised written and graphical format using a vocabulary of specific assessment terms. This standardised vocabulary was used to avoid ambiguity for the high-level reader, for whom such reports were usually to be written – the US Chief of Naval Operations in real life. Our conclusions and recommendations had to address the aircraft's compliance with the operational specification, its operational utility and acceptability, the level of mission accomplishment that was likely to be achievable in the aircraft's intended operational role, and the identification of any improvements that had to be incorporated before the aircraft could be released for service in the Fleet. Most of us had to put in at least twelve hours a day, six days a week to stay on top of the work required, which was, and is still, a common feature of all the test pilot schools.

In those days, there were no personal computers, no laptops, nor even pocket calculators, and so all our calculations were carried out with slide-rules and logarithm tables, and all our writing by typewriter or by hand. Flight data from our test manoeuvres were recorded either by photo-panel or on paper trace, which we then had to examine frame-by-frame or inch-by-inch to extract the parameters selected for analysis. These data would be supplemented by pencilled notes that we had made of observations in flight during our test manoeuvres, which included data from cockpit or hand-held instrumentation. We also used voice recorders to record our subjective descriptions of aircraft behaviour.

There are five main areas of interest in evaluating a new aircraft:

(1) its performance (including range and endurance; acceleration rate; maximum and minimum level speeds; climb, descent, glide, and turn rates; stall boundaries; take-off and landing distances etc);

(2) its flying qualities and its ground or deck handling qualities (i.e. how easy or difficult it is to control and manoeuvre throughout the required flight and ground-handling envelope, including recovery from stalls and departures);

(3) an assessment of how well it can perform its specified missions across the environmental range intended (including launch and recovery, mission radii, target tracking, formation and night flying, in-flight refuelling, etc);

(4) its aircraft systems (including flight-control characteristics, lift/ drag devices, hydraulics, electrics, fuel, cockpit environment, stores management, weapon release, etc); and

(5) its cockpit ergonomics and human interface (ingress and egress, cockpit instrumentation and displays, control and switchery operation, outside view, and weapons systems management etc).

Apologies to my former colleagues if I have forgotten anything – this is not intended to be a text book!

Fixed-wing test pilot students on USNTPS courses flew eight or nine different aircraft types; rotary-wing pilots, six. My course aircraft included: the A4B Skyhawk attack aircraft; three fighters: the F8K Crusader, F4J Phantom, and the F104 Starfighter (at Edwards AFB); the YOV1 Mohawk twin turbo-prop observation/attack aircraft; the T1A Seastar and T28 Trojan (already described); the S2 twin-piston Tracker anti-submarine warfare aircraft, the Lockheed QT-2PC 'Quiet' Reconnaissance aircraft, and the Calspan B-26 variable-stability demonstrator. In addition, I flew the X26A Schweizer glider, and the U6A Beaver and NU1 Otter single-piston aircraft on occasional weekends, towing gliders for a bit of light recreational relief.

First flights in any of the fast-jet types in particular were approached with especially careful preparation and a touch of anxiety. For my four-flight Naval Preliminary Evaluation exercise on the F4J Phantom, for example, I spent a lot of time mentally rehearsing my flight-test profile, sitting in the cockpit in the hangar. The day before my first flight, I even carried out an acceleration run along the runway in full reheat, terminating the run at the point of nose-wheel lift-off, just to get the feel of the machine without lifting her into the air. Even with all this preparation, the F4J's performance blew my socks off. On my first take-off and climb (clean configuration with full internal fuel), the aircraft went up like a rocket in full reheat. I hurtled up through 20,000ft in fifty seconds and passed through 35,000ft in less than two minutes, still climbing like a ding-bat. There was no instrumentation aboard this USN fleet aircraft, so all my data collection on the way up had to be noted by hand on my knee-board. When I examined my figures once I had finally caught my breath, I realised that I had written them down like an automaton, hardly aware of what I was doing. My brain had had a hard time keeping up, so all that rehearsal in the hangar had paid off. After four flights, I had mastered the brute; indeed I had fallen in love with it (I have to admit that I was an unashamed serial polygamist when it came to aeroplanes), and I had collected a ton of data and flown a hundred test manoeuvres – enough to write a final report that was nearly an inch thick!

All our reports had to contain a brief front page summary, a very concise abstract giving the short-of-time Naval Chief a quick answer to the following key questions:

(1) Does it (the aircraft or system) meet, or is it capable of meeting, operational requirements (subjectively and quantitatively) as specified?
(2) What MUST be corrected before entering operational service?
(3) What features might limit its operational capability and therefore warrant improvement (if possible/practical/affordable)?
(4) What are the unsatisfactory features that can be tolerated but should be avoided in any future design or development? And finally:
(5) What (if any) further testing needs to be carried out?

Here's a short clip from my abstract on the F-4J to give you a flavour (note the US spelling and use of standardised vocabulary):

> The F-4J airplane has the ability to perform the all-weather fighter mission. Exceptional maximum power turning performance at low altitude enhances its suitability as a fighter airplane. Poor transonic longitudinal flying qualities in maneuvering flight, objectionable accelerated stall characteristics, and poor sub-sonic turning performance at high altitude limit accomplishment of the all-weather fighter mission and should be corrected as soon as possible. Inadequate longitudinal trim rates in configuration PA, lightening of the longitudinal control forces as the stall was approached in configuration PA, and load-factor overshoots following sudden pull-ups were objectionable characteristics which should be avoided in future design. Further testing to evaluate the reliability of drag-chute deployment is recommended. Hot weather trials to evaluate the adequacy of the cockpit cooling system are also recommended.

The subsequent 115 pages of the report contained the detailed descriptions, analyses and data behind these and forty other conclusions and recommendations.

Speaking of flight-test reports, most of the twenty or so that we were required to produce during our eight-month course had to be typed and presented in a standardised NATC format. With her superb touch-typing skills, Caroline volunteered to become my part-time secretary for these exercises, turning my sometimes messy pencilled drafts into neat and tidy typed pages, and I'm sure that this must have helped my grades. Even so, my early reports were returned peppered in red ink underlining English spellings of words like manoeuvre (US – maneuver), centre (center), aeroplane (airplane) etc, of which there are quite a lot in flight test reports as you might expect. I managed to switch over to American spelling after a while, but Caroline assumed that my new American vocabulary was just poor (slightly dyslexic) spelling, and touch-typed it straight back into proper English. Eventually, my American tutors gave up!

In all, I flew 131 flying hours on fifteen different types of aircraft, of which over a hundred hours were flown as pilot in command. But all this flying was not without incident. During my time there, the test centre experienced a clutch of serious technical or maintenance incidents, and I experienced three, all traced back to questionable maintenance standards or procedures. It became such a concern across the whole of the Patuxent River airbase that all flying operations were brought to a halt for days while an in-depth safety review was conducted, and all technical procedures and practices thoroughly reviewed and overhauled. Some of the incidents were traced to such basic errors (e.g. missing split-pins, loose nuts, etc) that sabotage or mischief was listed as a possible cause, but this was never established beyond doubt.

Of my three maintenance-related incidents, two could have resulted in the loss of my aircraft – and one actually did …

The F-8K Crusader Incident

The first of my 'incidents' concerned a mechanical defect on an F8K Crusader, which I flew during the 'Flying Qualities' phase of my course. By way of background, the Crusader is a single-seat, single-engine, supersonic (Mach 1.90) carrier-based air-superiority aircraft – the last American fighter to be equipped with guns as the primary weapon – which earned it the title 'The Last of the Gunfighters'. It was notable for its variable incidence wing, which could be pivoted 7° upwards from the fuselage in high-lift configurations for take-off and landing. As far as the pilot was concerned, it felt more like lowering the fuselage rather than raising the wing, because the resulting attitude was more nose-down for a given airspeed, which thus improved the forward view on approach and also increased the tail-pipe-to-deck clearance on touch-down. At the F8's optimum approach airspeed – set relatively high to improve throttle-response and glide-path control – carriers often had to steam at full speed into wind to reduce the aircraft's relative deck-speed at touch-down. Despite this, it was still too easy inadvertently to allow a high sink rate to build up from which recovery was poor due to slow engine response at low RPM (remember my Vampire incident?). If the pilot was too slow in spotting a high sink rate developing, it could lead to a catastrophically hard landing. Indeed, the F-8 earned itself a reputation as an 'Ensign eliminator' when it was first introduced into service because it was so unforgiving in this respect. (An Ensign, by the way, is the lowest USN commissioned officer rank, not the ship's flag!) Landing the aircraft on a nice, long runway which doesn't move about like a carrier in a heavy sea, was much less fraught.

My incident flying the Crusader, however, had nothing to do with slow engine response (I had learned my lesson in the Vampire). Mine was due to a loose nut (of all things) – a loose idle-stop in the throttle track – which allowed the throttle to slide into the fuel cut-off position as I closed the throttle to the idle position on touch-down. It suddenly became awfully quiet as I raced along the runway with my RPM unwinding, but the residual hydraulics of my dying engine provided just enough brake-power to stop the aircraft before I ran out of tarmac. I trundled off the runway with the last of my momentum, but had to sit on the taxiway for quite a while before anyone realised that I was stuck. Unfortunately, the shut-down of my engine had also shut down my radio and so I was deaf and mute to the world. Like many USN aircraft of the time, the F-8 had no on-board batteries (which, I think, might have been to avoid carrying battery-acid at sea, with the possible hazard of it spilling in rough seas). This meant that as soon as engine power had been lost, all of the electrics failed too. The aircraft was equipped with a ram-air turbine for in-flight engine failures, but on the ground at low or zero airspeed, this was not much use. I must have sat on the taxiway drumming my fingers for half an hour, unable to call for help. Finally, the penny must have dropped in the air traffic control tower and a tug was sent out to tow me back to the TPS flight line. But the delay gave me time to contemplate the uncomfortable consequences that might have ensued if, instead of on touchdown, the engine had shut down as I broke into the landing pattern, which generally involved closing the throttle sharply to achieve a rapid deceleration to approach speed. A relight in those circumstances would have been impossible, and I would certainly have been forced to eject.

While awaiting rescue, I was also able to investigate the fault, which I traced to a loose retention screw and missing lock-wire on the throttle-stop. It was such a simple fault, but it could have resulted in the loss of the aircraft. I pointed this out to the maintenance chief when I got back to the flight line, but from his sideways look, it was clear that he was more inclined to suspect that I had run myself out of fuel rather than accept the slur on his maintenance standards. A subsequent check confirmed the fault I had reported, but it was not uncommon for a pilot's explanation of a technical abnormality to be mistrusted – as we shall see next.

The YOV-1 Mohawk Incident

The second incident involved a faulty nose-wheel anti-shimmy damper on the YOV-1 Mohawk, a twin turbo-prop, military observation and attack aircraft flown by the US Army and US Marine Corps. Ours was a prototype version of the aircraft, assigned to the TPS inventory after its development flight-trials had been concluded. We student test-pilots flew the aircraft on lateral-directional and asymmetric engine handling qualities assessments in the second phase of our course.

The incident occurred on landing at the end of a dual lateral-directional stability test-technique demonstration flight with my tutor, Royal Navy Lieutenant Commander Ian Normand. After he had finished his demonstrations and I had practised the techniques, Ian handed the aircraft over to me for the recovery back to base and sat back to enjoy the view of the Chesapeake waterline from his side of the cockpit. I entered the landing pattern and carried out a normal approach and touch-down, but at the point of lowering the nose-wheel onto the runway, a sudden and most violent shaking broke out that felt as if the aircraft was about to tear itself to bits. I wrestled with the aircraft to keep it straight and bring it to a halt on the runway, but when we got out to inspect the aircraft for damage, we saw that both main-wheel tyres had burst. Once again, I had to suffer the embarrassment of being brought back to TPS by tug, leaving the Mohawk and its smoking tyres to be doused by the attendant fire vehicle.

After an examination of the aircraft in the hangar, the maintenance crew's verdict was that I must have been a bit heavy on the brakes to have caused the tyres to burst, for no other fault had been discovered. Ian must have thought so too for he did not rally to my defence. I was pretty sure that I had not been heavy-footed and said as much, but I accepted their verdict with humility and had to put up with the scuttlebutt on the flight-line that the double tyre-burst was down to heavy British flying boots.

New wheels and tyres were duly fitted and the aircraft cleared again for flight a few days later. By coincidence, I was scheduled to fly it on another dual, this time with an American Army flight-test tutor who had flown the Mohawk operationally and knew the aircraft well. Aware of my tyre-bursting incident a few days before, he offered to demonstrate the landing at the end of our sortie (probably believing, like everyone else, that my landing technique must need correction). My instinct was to decline his offer as a matter of pride so I could prove that my technique was sound, but some inner voice suggested it might be prudent to let him have his way. I handed over the controls and watched him carefully as he performed the approach and landing. His touch down was, of course, perfect, and I saw the look of satisfaction pass across his face as the main-wheel tyres kissed the tarmac. But as he lowered the nose-wheel onto the ground, the same violent shaking occurred again, just as it had on my landing on the previous flight. Again both main-wheel tyres burst. After he had wrestled the aircraft to a stop I looked across at him with a raised eyebrow, and he looked back at me with a puzzled frown; we both suffered the embarrassment of being taken back to TPS sitting on the back of a tug, leaving our stricken Mohawk again sitting forlornly on the runway.

This time, a really thorough investigation took place, which revealed that the fault was in the nose-wheel shimmy-damping mechanism rather than the pilot's boots. It was another simple case of a nut and bolt working loose (a missing lock-wire again), which effectively allowed the damper to become disconnected. Amazingly, despite the investigation that had supposedly followed the first tyre-bursting incident, the fault had not been spotted by the maintenance crew. The nose-wheel assembly, freed of any restraint from the damper, had broken into a wild, dynamically unstable oscillation immediately on touch-down, and this had shaken the whole aircraft from side to side so violently that both tyres had been torn off their rims. As with the Crusader incident, here was another case of jumping to an easy conclusion – that it must have been mishandling by the pilot. In this complacent frame of mind, it had seemed unnecessary to look further into the first incident's cause. It was this sort of attitude, coupled with a degree of professional negligence, that had without doubt contributed to the Naval Air Station's poor incident record in the months before.

CHAPTER 8

While it is true that most aircraft accidents involve pilot errors of one sort or another (as I will discuss later), it is unwise to rush to that conclusion, especially if it excludes an examination of other possibilities. There are often also less-visible contributory causes – training, technical, ergonomic, weather, etc – each not individually serious enough to cause an accident, but adding up. An otherwise recoverable error by the pilot on top of a number of such contributory factors can be the straw that breaks the camel's back – and an accident ensues. The pilot may get the blame, but he is rarely entirely responsible.

Chapter 9

The A-4B Skyhawk Incident

The third and final incident during my time at Patuxent River was the most serious, and it was traced to a failed engine turbine blade in an A4B Skyhawk. The Douglas A-4 Skyhawk is a single-seat, single-engine, sub-sonic attack aircraft developed for the US Navy and US Marine Corps in the early 1950s. The aircraft had a distinctive delta wing, notably fitted with independent gravity-operated leading-edge slats, which gave the aircraft some pretty unpredictable handling characteristics during high-g manoeuvres when slats could deploy asymmetrically – which was a bit of a disadvantage when you were trying to track (or evade) an enemy fighter! The aircraft had a maximum take-off weight of only 24,500lbs, and yet it could carry a bomb load equivalent to that of a World War II B-17 Flying Fortress (8,000lbs). The A-4B models on the TPS inventory at Pax River were powered by the Wright J-65 turbojet, but later models (from the A-4E onwards) were fitted with the Pratt & Whitney J-52 engine. Skyhawks played key roles in the Vietnam War (US and Australian pilots), the Yom Kippur War (flown by the Israelis), and the Falklands War (flown by Argentinian pilots). More than sixty years after the aircraft's first flight in 1954, some of the 2,960 manufactured until production ceased in February 1979 are still flying today (in 2020).

I took off in my Douglas A4B Skyhawk (BuNo 142085) early one May morning to carry out a performance testing profile at high altitude over the 'Delmava' peninsula (short for the parts of Delaware, Maryland, and Virginia that the peninsula comprised), which formed the eastern shores of the Chesapeake Bay. The take-off and climb went without incident until about 20,000ft, when there was a sudden loud bang, the aircraft shuddered, and the fire warning light came on. I had just nosed out of the top of a 10,000ft-thick belt of heavy altocumulus cloud, and the clear blue sky above me beckoned. It would, however, be a fleeting glimpse. I shut down the engine immediately and switched off the fuel so as not to feed the fire. Like the Crusader, the Skyhawk was not fitted with batteries, and I knew therefore that the radio would fail as soon as the engine died. With RPM decaying rapidly, I put out an emergency call: 'Tea Kettle 085, Mayday, engine fire, turning back'; but before I could receive an acknowledgement, the radio went dead. Suddenly I was on my own, cut off from the outside world. I had eased the nose up to gain extra height while decelerating to my gliding speed, and now I commenced a descending turn onto a reciprocal heading, turning back in the direction from which I had come.

Settling into my glide, I was soon re-entering the thick dark cloud that I had exited just moments before. It felt somehow as if I were being swallowed up by it and I fought back an almost irresistible urge to hold my breath as I plunged below its opaque surface. As a contingency provision for just such an engine-out situation, the A4 was equipped with a ram-air turbine that deployed automatically as hydraulic pressure was lost (like the F-8 earlier described). This vital piece of equipment now came into its own, providing me with just enough hydraulic pressure to power my flying controls, and just enough electrical power to run my flight instruments (but not my radio). The fire warning light remained illuminated, but there was no other sign of fire that I could detect from the cockpit and so I rather hoped, with the fuel supply now cut off, that the fire had gone out. There was nothing more that I could do anyway even if it still raged. For some reason that I never really understood, the A4B was not equipped with a fire extinguisher.

Hoping for the best, I decided to ride the aircraft down through cloud. I could control it adequately and maintain attitude and direction with my instruments despite having no visual cues, but I had no working navigation equipment and so could only guess at where I might end up. I had noted the cloud base at around 10,000ft on the climb out, so I knew that there would be height to play with once I broke out beneath it. There was a slender chance, possibly wishful thinking on my part, that there might be an airfield within gliding range good enough for a dead-stick landing. I knew of one airfield on the peninsula that might be suitable at a push – the former civil field near Salisbury – but that was probably already too far East of where I was likely to emerge from cloud.

I held a steady course Westwards as my height slowly bled away. I knew that I could not possibly reach base on the other side of the Chesapeake, so I simply clung to the hope that there would be somewhere safe to put my aircraft down. I'd have been able to land my old Chipmunk in any farmer's field in these circumstances, but I'd need a mile of tarmac for the A4. Having abandoned an aircraft by way of ejection-seat once before, I was not especially keen to do it again. But without navigation equipment, I was steering purely by instinct; I could not even guess at what my options might be until I could see where I was. It was in this tangled state of mind that I became aware that tendrils of smoke had begun to infiltrate the cockpit. The smell of combustion soon reached my nostrils, and realising that the fire was still burning behind me, I snapped out of my reverie. Glancing at my altimeter, I saw the needles passing 15,000ft. The cloud was still thick and dark outside, and a feeling of doom and loneliness overtook me as I began to wonder whether I would make it after all. I ploughed on nevertheless, still holding my course and hoping for the miracle of finding a convenient landing strip where I could put this burning wreck down. Unknown to me, however, things were developing fast in the engine bay behind.

Suddenly, without warning, control seemed to slip from my fingers as the stick went slack; the hydraulic fluid that powered my flying controls was haemorrhaging from punctured pipes. It was this that had been feeding the flames; and now the precious fluid had all been lost. I felt myself lifted upwards in my seat straps as the aircraft bunted uncontrollably nose down, the elevators suddenly limp. The attitude indicator on my instrument panel, a small graduated black-and-white globe representing ground and sky, turned almost completely black as the aircraft plunged into a steep dive. The decision as to whether or not I should try to save my stricken aircraft seemed to have been made for me – there was nothing for me now but to eject. I reached up and pulled the ejection handle.

I knew what was coming. First, the canopy departed with the instantaneous roar of the 250mph slipstream that entered my topless cockpit. An instant later, I felt a mighty kick up the backside as the seat's cartridge fired. On the last occasion, escaping from my Hunter in Aden, I had used a Martin-Baker 2H ejection seat. That seat was an early model, propelled by a single, high-charge cartridge, which shot me out of my cockpit like a human cannonball at an acceleration rate of twenty-two times normal gravity. Every part of my body in that instant had suddenly weighed twenty-two times its normal weight, and as a consequence I had suffered a fractured lumbar vertebrae. But the seat on which I now rode out into the clouds was a more modern Douglas Escapac Mk1, fired initially by a cartridge half as powerful as the Hunter's, but then propelled upwards by a rocket, triggered by a lanyard as the seat rose up its guiding rails. Together these two boosts achieved the required ejection trajectory, but at a lower overall acceleration rate, which consequently did not inflict further injury to my spine. I had lost two centimetres off my height on my Hunter ejection and hoped not to lose any more! A moment later, the rough and tumble of seat separation and parachute deployment began, and I soon found myself suspended in my harness, still enveloped by cloud as thick as cotton wool. I had no sensation of falling; I seemed entirely cocooned, hanging limply in an ethereal state. Snow flakes rocked to and fro like crystal cradles in front of my face, floating earthwards at the same rate as I. I found myself captivated by the phenomenon and bemused by my unearthly predicament. But after a few minutes, things began to change. The cloud below me appeared to become thinner and less opaque, and the Earth's surface, strangely dark and featureless at first, began to take form. My altitude at this point must have been about 10,000ft, and through the misty remnants of cloud that encircled me, I searched for clues of where I might be. But there was nothing below my dangling feet than the dark and forbidding water of the Chesapeake Bay.

The two quick-release connectors that attached my harness to the parachute risers now sat at about the level of my ears. Above them, the webbing forked into fifty thin nylon suspension lines that splayed out tautly to the olive-green fabric of the inflated canopy. I remembered the casual manner in which I had clipped my harness to those connectors: a familiar, almost sub-conscious action as I strapped myself in to my doomed Skyhawk before starting up. These simple alloy spring clips, each with just a slight millimetre or so of purchase on the twin lugs of my harness, now held me suspended over such vertiginous emptiness that to quell the fear of falling, I felt compelled to loop my arms through the forks in the straps above my head.

The Bay's surface below me was closer now, ribbed by lines of little white-caps thrown up by an easterly wind; it looked cold and inhospitable. It was early May, and only a month before, the back-waters of the Chesapeake had still been frozen over. If I came down into the water, I knew it would be uncomfortable. But worse might befall me if I were not spotted, for in my single-seat dinghy I could be carried by the current out into the Atlantic, with little chance of surviving at all. I had been unable to communicate with anyone after the failure of my radio, and so no one would have known whether I had managed to eject; and if I had, where I might now be. But as my descent continued, the Eastern shore of the Bay appeared hazily in my view, a line of shallow breakers on a grey shoreline, indistinct at first, then clearer as the wisps of cloud dispersed. I studied it assiduously for a while, desperate to persuade myself that I was drifting towards it. The closer I got,

I thought, the better chance there would be of being seen. It still seemed a long way off, however, and with no effective control over my direction, I was entirely at the mercy of the wind.

Then, to the north of me, perhaps ten miles or so, I spotted an aircraft circling at about my height. It was only a distant speck, but I could make out the distinctive long lines of an F-8 Crusader (I had better eyesight in those days). I remembered then that my fellow student test pilot, Dale Iverson, had taken off immediately before me to carry out a similar high-altitude test exercise. I had watched him take off – his thundering reheat had made my diminutive A4 shake as I waited at the take-off point behind him. With his reheated thrust, he would have rocketed skywards and would have been well above the cloud-tops when my emergency occurred. He told me later that he had heard my mayday call and had immediately descended beneath the cloud to wait for me to appear. As I dangled unseen in my parachute, he was already searching for me. I waved my arms frantically at the circling speck knowing that it was a futile gesture, but somehow it gave me the feeling that I was not alone after all.

My descent continued as more anxious minutes passed. The detail of the shoreline, with all its muddy inlets and creeks, was now becoming very distinct, and I was soon certain that I was to be spared the watery end that I had feared. With my continuing eastward drift, I could see now that I would land a little distance inland from the beach, just missing a narrow strip of tall conifers that lined the shore. At about 1,000ft, I could estimate with reasonable accuracy that my descent path would end near to some farm buildings on otherwise open ground. Until this point, my descent had taken what seemed like twenty or thirty minutes, proceeding at such an apparently leisurely pace that I had had plenty of time to assess my progress and muse at my fate; but now it seemed as if I was dropping like a stone. Suddenly, the ground rushed up to meet me like an express train at full pelt. I hit it hard, landing in an ungainly heap about 20 yards from the front door of the clap-board, single-storey farmhouse that I had seen from above, and flattening a good-looking crop of onion stalks in the tumble that ensued.

The earth was wet from recent rain, and in collecting up my parachute and equipment, I churned my landing spot into a quagmire. It was still only about eight o'clock in the morning when I introduced myself to the inhabitants of the farmhouse – two elderly spinsters, just then in the middle of their breakfast. The sight that met them on their doorstep must have been alarming for I was caked from head to toe in the muddy residue of their precious onion patch. Wearing my drab-green military paraphernalia, my helmet and oxygen mask dangling from my arm, and my parachute and harness draped over my shoulder, I must have looked like the advance guard of an alien invasion! My English accent, foreign to their Southern-Maryland ears, might well have compounded this impression. The two ladies just stood at their door and gaped wide-eyed at me as I apologised for my rude intrusion; but their shock soon turned to excitement as they realised what celebrities they would now surely become. I was invited in for a cup of tea – as befitted the arrival of an Englishman – and the couple were soon hanging on my every word as I recounted the story of my escape. And more excitement was yet to come after I had made a telephone call to TPS to tell them where I was. A search and rescue helicopter arrived about thirty minutes later to pick me up, landing on the same muddy patch of ground on which I had inelegantly alighted an hour or so before, and flattening any of the onions stalks that I had left standing!

Captain 'Rube' Pritchard USN, Director of the School, greeted me like a long-lost son when I arrived back in the TPS operations room. Dale Iverson in his circling F-8 Crusader had watched my aircraft streak out of the clouds in flames and plummet into the water, and had then searched the scene vainly until low fuel had forced him to return to base. Not noticing me hanging in my parachute some ten miles to the south and seeing no sign of my escape, he had radioed 'no parachute, no dinghy'. Everyone at TPS had thus come to the conclusion that I must have perished with my aircraft, and the CO was preparing himself to convey some bad news to Caroline. My telephone call from the farmhouse, therefore, had relieved him of that onerous duty, but by then the local radio had got wind of a 'missing English pilot' and had broadcast the breaking news before he or I were able to ring Caroline at home.

My Skyhawk was never found, and when I think back on the event, I still picture it lying buried in the soft mud at the bottom of a Chesapeake creek, just as I left it fifty years ago, my stop-watch still ticking in its holder, my knee-board still in the cockpit. More likely, the aircraft broke apart on impact and so may not be in quite the state I fondly imagine it. The USN sent a salvage vessel to search for it with sonar, but could not find any wreckage, even though its position had been roughly plotted by my circling friend. The crash investigators could therefore not be absolutely certain of the cause of the fire, but concluded that it must have been the result of a fracturing turbine blade. It appears that such failures were not uncommon on the J-65 engine and that previous events of the kind had had similar outcomes. The fragmenting blade, they surmised, had ruptured the engine casing, thus allowing a white-hot jet of burning fuel/air mixture to ignite any combustible material that it touched. The flames of the fire would have been fanned to a very high temperature by the airflow, and could easily have reached temperatures high enough to ignite the hydraulic fluid and melt the alloy control rods. This explanation, at least, accounts for the loss of flying control that occurred late in the aircraft's descent.

Following my first use of a Hunter 'bang' seat in Aden, my back injury took four weeks of healing and physio-therapy and a month or so of convalescence before I was declared fit enough to fly again. By contrast, after using a rocket-assisted seat to escape from my Skyhawk that early morning, I was at my desk at lectures that very afternoon and airborne again first thing the following day. The development of rocket-assisted ejection seats has practically eliminated the ejection injuries that were common-place before. If I had needed to evade capture after my first ejection in the desert, I would not have been able to run very far before seizing up. Following the second, however, I could have run for miles!

I graduated from TPS with the US Navy League Award as the outstanding student test pilot of my class. As they say in politics, any publicity, sometimes even bad publicity, can be helpful in one's ascent up the professional ladder because it gets you noticed, while the reasons for your fame (or notoriety) are soon forgotten. I sometimes wonder, therefore, if it really was my exceptional skills as a budding test pilot that brought my name to the fore, or the number of times that my name must have featured in NATC incident reports!

Chapter 10

Probe Probare – (Properly to test) – Inscription on the crest of the Aeroplane and Armament Experimental Establishment, Boscombe Down.

Returning to the UK, the late-October dawn was cold, grey, and misty as we were driven through the Cotswold countryside. We had just arrived at RAF Brize Norton after the overnight 'red-eye' VC10 flight from Washington en route to Boscombe Down in Wiltshire, the Aeroplane and Armament Experimental Establishment (A&AEE) – my next posting.

At the time of my return to the UK in October 1970, British aviation research and development fell within the bailiwick of the Ministry of Technology (Aviation Supply), whose political head was then the Rt. Hon. Tony Benn. His government department oversaw (inter-alia) the activities of the A&AEE at Boscombe Down, the Royal Radar Establishment at Malvern and Pershore, and the Royal Aircraft Establishment, based principally at Farnborough and Bedford (Thurleigh) with other testing facilities at Llanbedr and West Freugh. If I had been given the choice of postings as a newly graduating test pilot, I would probably have opted for Aero' Flight at RAE Bedford because some very interesting experimental prototypes were still being flown there – even some of those weird and wonderful flying machines that had featured in the Eagle magazines and aviation press of my growing up. In the 1950s and '60s, these experimental machines had included the supersonic research Bristol 188 and Fairey Delta II, the latter type famously flown by Peter Twiss, the first pilot to exceed 1,000mph in level flight in 1956. Other prototypes flown at Bedford during that period had included the Short SC1 VTOL research aircraft, the Hunting H-126 'Jet-Flap' aircraft built to investigate the use of 'blown' boundary-layer control, and the P 1127/Kestral FGA1, the latter destined eventually to evolve into the Harrier. I think I rather immodestly pictured myself, like Chuck Yeager in 'The Right Stuff', pushing back the frontiers of science in these exotic machines!

As a rookie test pilot then only 26 years old, however, I had not yet appreciated that things in the experimental field had changed a lot, at least in the UK – that our glorious past was no predictor of our future. By the late 1960s, UK government-funded development activity was already running out of steam (and resources). In the twenty-five years that had followed WWII, the UK had spent itself out in a vain attempt to keep up with the USA and the USSR in aviation design and development. It was this quarter-century of government investment that had spawned the many different aviation research projects and prototypes that had moved UK technology on in leaps and bounds. However, while the scientific and engineering benefits of all that research had kept the UK in the top league of aviation nations, the financial cost had been too high for a relatively small economy such as ours. We had kept up in terms of defence capability, punching above our weight internationally and within NATO, but too many UK projects had failed to provide the return on investment that would have justified the expense to an increasingly critical exchequer. National debt was growing out of proportion to our means, resources were stretched, and priorities were shifting. Pressure to cut our cloth to suit our means was building fast.

In that last 'golden' quarter-century of aviation development, the RAE had been at the forefront of aerodynamic research, with RAE Bedford at the centre of design and innovation, focusing on priority areas such as supersonic flight, blind-landing, lift augmentation, vertical take-off and landing, and augmented stability and control. Its wind-tunnels, its aerodynamic modelling, its prototypes, its test-beds and flight simulators were the hardware that tested the science and put theory into practice. By 1970, however, so much of what had been learned from those decades of research was already being incorporated into future designs. There were still a few important projects on the experimental to-do list, but Bedford's heyday was over; the site of so much invention, creativity, and experimentation was already in decline.

So instead of flying the Fairey Delta II and the experimental prototypes that I had coveted, I would be testing and evaluating new aircraft, weapon systems and other new equipment at Boscombe Down, carrying out acceptance and proving trials prior to their release to the front-line services. While some experimental and development work was also carried out at A&AEE, for example on new in-flight refuelling and weapon systems, Boscombe's main role as the MoD's aviation quality controller, was to check that all new equipment entering service actually did what it was required to do, rather than simply take the manufacturer's word for it.

To carry out this function effectively, A&AEE employed several thousand staff of which only some 200 were service personnel, the majority of the latter being aircrew or armament officers. Flying Division incorporated five test-flying squadrons: 'A' (Fighter & Trainer aircraft); 'B' (Bomber and Maritime aircraft), 'C' (Naval aircraft), 'D' (Rotary-wing aircraft); and 'E' (Transport and Utility aircraft). The Empire Test Pilots' School (ETPS) was also part of Flying Division (yes, even now, in 2020, we still call it that, though our 'glorious' Empire is long gone), as was the meteorological office, the fire & rescue service, and air traffic control.

Test pilots and flight-test aircrew of the test squadrons (and ETPS) were serving military officers and NCOs from their respective services, supported by the civilian engineers and technicians of Technical Services Division who carried out all aircraft maintenance and repair. Trials were devised and managed by the trials divisions in the four discrete specialisms: aircraft performance & handling-qualities; armament and stores-release; navigation, radio, and weapon systems; and engineering and environmental systems. These divisions were supported by photographic and technical services divisions, and a computer, telemetry, and instrumentation group. Each division was headed by a group captain or civil-service superintendent of equivalent rank, and the Establishment as a whole was directed by an RAF Air Commodore and a civil-service Chief Superintendent.

I would be privileged to serve at the Establishment for four test-flying tours between 1970 and 1988 at levels from flight lieutenant to group captain and so got to know it, its personalities, and its activities, quite well. The base was remarkable for its specialists in every field and for its range of testing facilities. For example, in addition to its test-flying squadrons and its flight-testing divisions, Boscombe was also the home of:

- an aeronautical engineering apprentice school;
- an EMC and EMP testing facility (electro-magnetic compatibility and pulse);
- an aviation medicine centre;
- an airborne-forces parachute testing unit;
- a high-speed blower tunnel to test aircraft canopies and ejection seats;
- an environmental hangar which could expose whole aircraft to temperatures between plus and minus forty degrees centigrade;
- and the RAF Handling Squadron, which wrote or edited the aircrew operating manuals for new aircraft and equipment entering service.

From the 1950s to the end of the '80s, Boscombe Down's facilities for aircraft and systems development and testing were probably second only in the world to those of the USA. It was a heady time to be involved with flight testing.

On first seeing Boscombe's vast concrete aircraft dispersal area, I was struck by the huge number and variety of aircraft types, there being hardly a parking spot unallocated between one end of the half-mile-long pan to the other. The air throbbed, whined, or roared with engine noises and wreaked with the acrid fumes of starter-cartridges and turbine-engine exhausts (for which I would become quite nostalgic in later life). I counted examples of practically every aircraft type currently in or about to enter RAF, RN, or Army service, plus a few older aircraft such as the Comet (for navigation and radio trials), several Hunters (for ETPS and pilot-continuation training, and work with the aero-medical centre), a scarlet-and-white Gloster Javelin (for pressure error testing and calibration), a fleet of Sea Vixens (for missile trials), several marques of Canberra (for training, and for ejection-seat, target-towing, and icing trials), a Bristol Britannia (for overseas trials support), three Harvards (for slow-speed photo-chase of parachute trials), a couple of Bassets (for executive transport) and even a Hastings and Beverley (for parachute testing). The range of current in-service or pre-release aircraft spanned: Harrier, Jaguar, Lightning, Phantom, and Buccaneer at the fast-jet end of the spectrum, to transport aircraft like Andover, Argosy, Belfast, and Hercules. Bombers and maritime aircraft then under test included Canberras PR9 and TT18, Victor, Vulcan, Nimrod and Shackleton. ETPS flew aircraft from the training fleet plus a two-seat Lightning, a Viscount, two Bassets (one eventually modified as a variable stability training aircraft), a Twin Pioneer, and a Jet Provost. The rotary-wing squadron, on the southern side of the runway, operated Wessex, Scout, Whirlwind, Lynx, and Puma. It would have been any aviation enthusiast's dream to walk that line. Moreover, all these aircraft were engaged in active trials flying or training, not laid out like some air-show static display. It was a noisy, bustling, and exciting place to enter as a young test pilot as eager as I was to fly as many of these aircraft as I could lay my hands on (which, by and large, I eventually did).

I arrived on 'A' (Fighter test) Squadron in November 1970 to join Squadron Leaders Tom Lecky-Thomson and Graham Williams who were still basking in the glory of their recent trans-Atlantic record flights in Harrier and Phantom respectively (the Harrier taking off famously from St Pancras station). Graham had been my first flight commander in Aden and was now senior test pilot of 'A' Squadron at the time I joined it. Wing Commander Ian Keppie was the CO, Geoff Thirkell was its maintenance chief. It was a small and select group of fast-jet flyers that, with my arrival and that of a French test-pilot-school graduate, Jim Hawkins, comprised ten test pilots in all, complemented by two specialist trials navigators. In my first few months on the Squadron, I flew the Phantom F4K and F4M and the distinctive scarlet-and-white Javelin pacer aircraft. I was just about to join the Harrier trials programme when other events intervened.

The Multi-role Combat Aircraft

It had been decided somewhere in the upper echelons of the Establishment that the Multi Role Combat Aircraft (MRCA – later to be named Tornado) would be allocated to the bomber test squadron ('B' Squadron) during its design and development phase. It was to be primarily a bomber after all, albeit a very fast and capable one that could have come straight out of the new-generation 'fighter/ground attack' mould. Wing Commander Ned Frith, a former Canberra and Vulcan bomber pilot and CO of 'B' Squadron at the time, had recently taken on the role of first A&AEE MRCA project test pilot and was already engaged in the aircraft's cockpit and displays design, working alongside his opposite numbers from the German and Italian official flight test centres. As a former bomber pilot working with German and Italian fighter pilots (who could talk the talk and walk the walk), Ned had come to the conclusion, for the sake of UK influence in the early stages of MRCA's cockpit design, that a British fast-jet test pilot should join this international group. At that time, all the 'B' Squadron test pilots were former 'V'-bomber, Shackleton or Canberra pilots, and there was no fighter/ground attack experience to draw upon. I am guessing, therefore, that there must have been a conversation between the Superintendent of Flying Division and the COs of 'A' and 'B' squadrons to decide who should be transferred from A to B to fill the new vacancy. To cut a long story short, and for reasons that I was never privy to, I was the one chosen.

This proposed move did not please me at all, however, because the MRCA in 1971 was still literally at the plywood and cardboard mock-up stage, and so the prospect of my actually flying it seemed remote (indeed, it was not until four years later that I became the first RAF pilot to lift its wheels into the air). Moreover, after the cancellations of the TSR2 and F-111 as possible contenders for the post V-bomber generation of RAF interdiction-strike aircraft, MRCA seemed vulnerable to suffer the same fate. With this in mind, my transfer to 'B' Squadron felt like a kick in the teeth, which quite bruised my fighter-pilot's fragile ego at the time. It felt a bit like being sacked from the Formula 1 racing team and relegated to driving HGVs instead! At first, I wondered how I might wriggle out of it and protested (rather lamely) to my boss, but I could see the logic of the transfer and so buttoned my lip and put a brave face on it.

The following day, I reported to my new squadron and started my assimilation into a very different culture than I had previously been used to. It took me some time to adapt. 'B' Squadron was a much larger squadron than 'A', with pilots, navigators, air electronics officers, flight engineers and load-masters, who made up the crews of the large bomber and maritime aircraft then being flown. Pilots of these aircraft, unlike single-seater fighter or attack pilots, were not free spirits able to act independently; they were aircraft commanders who had crews to manage. Mature, steady leadership and good crew management were as important as flying skills for the success of any large-aircraft bomber or maritime mission. With most of my flying carried out solo until then, my single-seat fighter aircraft had felt like an extension of my own being, which I could throw around the skies almost entirely uninhibited. But now I had other aircrew sitting behind me who needed to be informed, considered, and consulted – as well as depended upon. Actually, it felt more grown up to be the captain of a crew, and I soon realised the operational benefits of having a colleague (sometimes more than one) with whom to share the flight-operations and navigation tasks.

Low-level or complicated operational or trials missions in poor weather or at night would have been beyond the capacity of most single-seater pilots with the technologies then available. Moreover, after the two engine failures and bale-outs that I had suffered in my single-engine flying career so far, it was also comforting to have more than one power-plant. There was a surprising additional benefit too. As a single-seat pilot used to downing a hasty coffee and corned-beef sandwich between flights, I was rather taken aback to find my new squadron equipped with a kitchen and a dining room (and a catering warrant-officer too) for post-flight meals and the preparation of in-flight rations. This facility, it appeared, was standard on large, multi-engine aircraft squadrons, and so 'B' Squadron aircrew expected to have one too. Clearly, flying the 'V' bombers, Nimrods and Shackletons then being tested at A&AEE was hungry work. Like an army on the march, it seemed that this part of the Royal Air Force flew on its stomach, and so occasionally I too was treated to a proper lunch. I soon began to think it was not at all a bad thing to have joined 'B' Squadron; and its CO, Ned Frith, was a charismatic and inspiring boss, who was good company and very good to work with.

I worked on MRCA cockpit layout and electronic displays design with my German and Italian opposite numbers between 1971 and 1975. This required a monthly trek up to the British Aircraft Corporation base at Warton in Lancashire for meetings, mock-up assessments and simulation sessions. Pilots, rear-crew, and engineers from UK, Germany, and Italy made up what was called the CCCC – the cockpit control and coordination committee. With three nations' aircrew and their air staffs to satisfy, every detail of design took time to develop to the point of agreement. These were large meetings with up to thirty attendees, all with different backgrounds and interests, and all with their own ideas – which was why the whole thing took four years or so to complete. Having said that, the process was remarkably collegiate and was assisted considerably by BAC's project test pilot, Dave Eagles, and specialist navigator, Ray Woolett. Our work included, for example, the design of each individual cockpit instrument, and each electronic display and system control panel. Important too was how these pieces of equipment interacted with each other and with other aircraft systems, and their layout and disposition in the two cockpits to balance the work-load between front and rear crew. We also had to debate and agree how all the switches and buttons should operate, the illumination levels of displays and captions for the extremes of direct sunlight and at night, and even the font and case of the labelling for clarity. Our objective from the beginning was to make the MRCA/Tornado's two cockpits as effective and as ergonomically efficient as we could so as to ensure that the aircraft's complex systems could perform at their technological best, unimpeded by unnecessary, cumbersome, or error-prone operating routines. The new aircraft would represent cutting-edge technological innovation for the 1970s, with automatic terrain-following, computerised fly-by-wire controls and stability augmentation, swing-wings for Mach 2+ performance, reverse thrust for short landings, the latest and highest spec navigation and ground-mapping radar equipment, and the most modern head-up and head-down cockpit displays then available. To be involved with its development as a young test pilot was a privilege.

Cockpit Design

The cockpit designs of earlier aircraft – the layout of the knobs, levers, and switches, and the design of displays and instruments – was more often driven by expediency rather than ergonomic sensibility. Cockpits could be dark holes where things were bolted into awkward corners wherever space permitted, sometimes as an afterthought and sometimes fixed in such inaccessible places that controls could be reached only by feel or with rubber-like contortions of the wrist. Moreover, one aircraft's cockpit layout and instrumentation would be completely different to another, even in aircraft of similar role and function, there being little if any standardisation between aircraft type, armed service, or nation. If you flew foreign-bought aircraft, cockpits would feel idiosyncratic, and sometimes even the operating sense of displays and controls would be contrary. It is true that cockpit design in aircraft developed in the '50s and '60s was generally constrained by space or limited by technology or materials, but more often than not these poor layouts were the product of engineering expediency and insufficient thought about the human operator in the controlling loop. The sciences of human-factors and ergonomics was barely on the curriculum then, let alone developed. When you look at some of these older layouts, they seem agricultural by today's standards. And poor cockpit or display design will undoubtedly limit operational effectiveness and increase the operator error rate. In the worst cases, these poor designs have cost lives.

Pre-'70s designs were largely electrical/mechanical circular instruments, latterly including the head-up displays and projected map displays of the Harrier and Jaguar. By contrast, the F-111 employed vertical thermometer-like strip scales to represent airspeed, altitude, and vertical speed placed either side of a huge attitude globe. When I first flew the F-111, I found the instrument display mind-boggling because the strips often moved in contrary (and confusing) directions. The OR946 integrated cockpit display of the Lightning and Buccaneer was better; this display used clock-face instruments with moving needles for altitude, airspeed, and vertical speed, but a horizontal strip scale for airspeed and Mach-number. Tellingly, a conventional clock-face airspeed indicator had to be retro-fitted on the cockpit combing of the Buccaneer because it was easier to read and react to than the head-down horizontal strip.

Most pilots will be able to adapt to different methods of data presentation, but human adaptation can slow down mental processing. In any fast-moving operational situation (like flying fast-jets, for example), data and information displays must be designed to convey important flight and tactical information as efficiently as possible into the operator's brain (pilot, navigator, or other display operators) to enhance mental processing and therefore speed up decision-making. Designing the manner in which data are presented and conveyed, therefore, needs an understanding of how best the human brain assimilates and processes information of different kinds, and which human communication channels are best to convey it (sight, sound, feel, etc) across the interface between operator and machine. In the MRCA/Tornado development programme in those first four or five years of the 1970s, huge amounts of development and trials work was carried out by our cockpit design group – test pilots and specialist navigators working with systems design and human-factors engineers on every aspect of cockpit and display design. And much use was made of simulation to develop weapon-system controls and displays to optimise crew efficiency and to balance workload between pilot and navigator. As leaders in this field at that time, we also joined groups developing international standards across the industry.

Display technology and computer power has moved on hugely since those days, allowing a much more innovative approach to cockpit and display design. Modern cockpits like the F-35 Lightning, for example, make extensive use of multi-function flat-screen electronic displays with data and information projected onto the pilot's external field of view. There are no electrical/mechanical instruments, even as back-up. In addition, the pilot is equipped with a helmet-mounted display which combines flight and weapon-aiming data with night-vision imagery.

I have added this bit of detail to illustrate that behind every modern cockpit layout and display there is a great deal of 'human-factors' development work to be done. This invariably involves a lot of experimenting and trial and error before a design is passed as fit for purpose and ready for production. I also wanted to illustrate that the modern test-pilot's (and test operator's) work is often as much to do with human-factors engineering – making the operator's interface with the aircraft and its systems as seamless, intuitive, and efficient as possible – as deeds of derring-do and frightening themselves in the air!

'B' Squadron Flying

B Squadron's inventory at the time included a range of heavy aircraft, which I had the opportunity to fly occasionally as co-pilot, but most of my flying continued to be on the faster, twin-engine, twin-crew, types. For example, I flew hundreds of trials flights on the Buccaneer S2, originally purely a Royal Navy maritime strike aircraft later transferred to the RAF for the overland strike/attack role. I also joined 'C' Squadron's Sea Vixen flight, flying Sea Vixens Mks 1 and 2 for the Martel TV-guided missile trials (more on these later). These were my principal flight testing projects during that period, but I also contributed from time to time to flight trials on various marques of Canberra. For example, I was once dispatched to a distant part of the world on a special high-level reconnaissance mission flying my favourite Canberra marque, the PR9, equipped with a high-definition camera on trial for satellite use. We flew at such a high altitude that my navigator and I were equipped with pressure suits and full-face helmets to protect us from the extremely low air pressures of the stratosphere should a cabin pressurisation failure occur. Our flight profile took us to an altitude just short of 70,000ft where the sky above us darkened to a deep shade of navy blue and the curvature of the earth was clearly visible. The experience was the closest I ever came to being an astronaut in low orbit. The mission was shrouded in secrecy and even now I am not able to talk about it in any detail, but at the time, I felt a bit like Gary Powers must have felt in his U2. Fortunately, I was not to suffer the same fate.

Having used an ejection seat twice in anger as already related, it was also interesting for me to become involved in the testing of new ejection seats from time to time – not using them myself, thankfully, but strapping our resident anthropomorphic dummy, 'P/O Phredd', into the hot seat, and ejecting him from our test-bed Canberra WJ638 over a range of speeds and heights. I also flew the hybrid Canberra Mk2/8, WV787, which had been extensively modified to become Boscombe's 'icing tanker' for the testing of fixed-wing aircraft anti-icing and de-icing systems. This modification included the fitting of a 600-gallon stainless steel water tank in the bomb bay, and the conversion of one of its fuselage fuel tanks to carry a further 300-gallons of water. The water was sprayed at pressure from a twenty-two-nozzle calibrated spray rake mounted on a boom behind the fuselage to deliver a finely-atomised water spray that could produce localised icing conditions. I flew it quite often to test the anti-icing and de-icing systems of several new aircraft types, but the icing trials that I best remember were those with Concorde. The reflection of its sleek anatomy in my rear-view mirror as it came up close behind me, its nose drooped like a preying mantis poised to strike, is an image that will never leave me.

The twin Avon-engine English Electric Canberra, first flown in 1949, was a remarkably versatile aircraft, as the above examples of its employment might illustrate. Moreover, fitted with four Hispano 20mm cannon (as it was on some of the foreign-bought versions), it also made a passable ground attack aircraft as well as an effective medium bomber. Conversion-to-type took place in the trainer version, the T-4, with a second pilot's ejection seat shoe-horned into the cockpit (with a clumsy hinged arrangement), which made it feel cramped and claustrophobic. The pilot's view from most marques of Canberra (and especially the T-4) was poor. It looked to me as if the original low-profile bubble canopy was a reluctant design concession in an otherwise aerodynamically sleek airframe – almost an afterthought, it seemed – the designers belatedly thinking that they had better let the pilot see where he was going! Later marques, some PR versions and the US-procured version, the B-57, gave the pilot a much better elevated view from a tear-drop blister-shaped canopy. This was great for the pilot but, except in the B-57 version, the navigator was still buried in the fuselage with practically no view at all.

Despite the Canberra's poor cockpit design, reminiscent of an old combine harvester, with knobs, switches, and levers scattered all over the place, I developed a special liking for the aircraft. It had a good turn of speed and a huge wing area that gave it excellent turning and high-altitude performance. On the other hand, even in mild turbulence, its low wing-loading gave the crew a very uncomfortable ride at low level. It was as well that we all wore 'bone-domes' (hard helmets), for our heads took a bit of a bashing on the canopy when flying fast at low altitude on a hot day, when atmospheric thermal activity was at its most energetic.

Because the Canberra's two engines were so widely separated, single-engine handling (flying with one engine inoperative or at idle power) could make the aircraft a bit of a handful at low airspeed. Indeed, the Canberra's asymmetric thrust in that condition made it a potential killer (like the Meteor, which had a similar engine configuration). I use the word 'potential' advisedly, because flying on one engine was not intrinsically a problem at all if you handled the aircraft with respect. Most aircraft will have a vicious side if you don't obey its simple rules; and if you didn't obey the Canberra's rules, it could become very vicious indeed (like a woman scorned, as the saying goes). Most Canberra fatal accidents (and there were quite a few) occurred when attempting a go-around from a single-engine approach, opening up the live engine to full power when the airspeed was below about 150 knots and the gear and flaps were extended. The rudder power (the aerodynamic control moment produced by the rudder at full deflection) at such low airspeeds was simply not enough to counteract the full asymmetric yawing force produced. Taken beyond this limit, the aircraft would then yaw and roll out of control into the ground so quickly that there would be very little time to eject. The rule was simple: once your airspeed was below 150 knots on a single-engine approach, you should consider yourself committed to a landing. Indeed, you would be putting yourself in grave danger if you tried to abort the approach at this late stage. The tragedy is that most, if not all, fatal accidents of this sort in the aircraft arose when practising simulated single-engine go-arounds, despite such situations rarely occurring for real.

As well as flying many different marques of Canberra, I also flew as co-pilot in the larger aircraft too, like Shackleton, Vulcan and Victor, all still undergoing trials of one sort or another in the early seventies. The Shackleton in particular made a big impression on me (not least on my eardrums), and it was so similar an aircraft to my father's wartime Lancaster B1 that it was easy to picture him in my place. Like him, however, even though its counter-rotating propellers reduced the ground-looping tendency, I was never allowed to land it. But I did qualify as captain of the four-engine Nimrod (the maritime patrol version of the de Havilland Comet airliner), and dropped sonar buoys and torpedoes (dummies, of course) on cooperative submarines in Lyme Bay. All this varied experience was quite unusual for someone with a single-seat fighter background, and I realised how lucky I had been to transfer to 'B' Squadron after all. I was certainly getting the best of both worlds – and making the most of it. It was here that Dave Foster, the Squadron's senior test pilot at the time, taught me his 'rule of five', which I have tried to follow ever since when making any decision to do with aircraft operations. This was not to respond or react instantly to a problem because things are bound to change, but, depending upon the situation faced, to allow at least five seconds, five minutes, five hours, or even five days to pass in due consideration before bursting into action. It rules my home life too and sometimes drives my wife potty.

The Buccaneer S2

In July 1973, I flew Buccaneer XV350 to the US Marine Corps Air Station Yuma in Arizona for hot-weather trials, accompanied by Peter Wilde as my flight-test observer and navigator. We were followed by the A&AEE Britannia carrying our trials officers and support team, numbering some twenty-four servicing and trials personnel in all. Our route to Yuma from Boscombe Down was via RAF Saint Mawgan (to top up fuel), the USAF base at Lajes in the Azores, the RCAF airbase at Goose Bay in northern Canada, and the USAF air base at Offutt near Oklahoma City. It was quite an epic journey, which would take four days in all. In preparing for the flight, I was concerned that I would be completely dependent on receiving Peter's navigation instructions over the intercom. We would spend seven long hours over the Atlantic on our route, and he had the long-range radio and all the navigation equipment in the rear cockpit. Should our inter-communications fail, therefore, there would have been no easy way for vital navigation information to have been passed between us. The two cockpits in tandem were effectively screened from each other (except for a narrow gap of about 3 inches) by a wind-shield designed to protect the rear crew from blast or debris should the canopy be shattered by bird strike (or shrapnel). Shouting between the cockpits, therefore, was not a viable solution; besides, removing your oxygen mask at high altitude could risk unconsciousness, which would not have been very helpful. I came up with a simple solution that put my mind at rest, and had a strong crocodile clip welded to a telescopic radio aerial long enough to pass hand-written notes between the two cockpits. In the end, my invention proved unnecessary for its planned purpose, but it was extremely useful in supplying me with Mars bars from the goody-box of rations stowed under the navigator's console (there was no room in the front). He could also have used it to tap my helmet should I fall asleep!

Peter and I spent six weeks in Yuma and flew thirty-seven trials flights before our return to the UK. Our objective in Yuma was to establish the Buccaneer's performance and handling limits at heavy weights and at high ambient air temperatures. In particular, we needed to establish the aircraft's flying characteristics at high angle-of-attack (AoA) in high all-up-weight configurations following single engine failures. From these tests, our results would enable us to make recommendations on safe flight-control limits and give advice on the handling techniques when flying the aircraft in such conditions. Dennis Sharp was our senior trials officer, and the two of us devised a test programme that included scores of single-engine decelerations to high AoA, and simulated engine failures on practically every take-off. These tests were carried out at progressively increasing all-up weight (AUW) and ambient air-temperature.

Depending upon fuel carried, the Buccaneer S2B could carry up to sixteen 1,000lb bombs, four in its rotating bomb bay and up to twelve more on its wing pylons. To reach the maximum take-off weight of 60,000lbs, the all-up weight limit we achieved at Yuma at 35°C ambient air temperature, XV350 carried a full fuel load of 20,000lb with 10,000lb armament (round figures). At 40°C, maximum take-off weight had to be reduced to 55,000lb for safety. Air temperatures higher than 35°C reduced engine thrust to the point where a safe fly-away at AUWs above 55,000lb AUW following an engine failure immediately after take-off was not possible.

Interestingly, at these high ambient temperatures and aircraft weights, the monitoring of tyre temperature became safety-critical. Tyre-bead temperature increased rapidly with aircraft movement due to the flexing of the tyre walls with wheel rotation. The several miles of taxiing required to reach the take-off point added up to fifty degrees centigrade to the tyre temperature from an already high starting point. This monitoring became really critical when multiple sorties were flown back-to-back during the day, as very little cooling actually took place in flight, or indeed between flights when the aircraft sat on hot concrete exposed to direct sunlight during refuelling and turn-around checks. By plotting tyre temperature throughout a typical working day, we saw that the rise was cumulative and unrelenting, and any significant use of the wheel brakes during taxiing or landing made it worse. Having neither reverse-thrust nor a tail-chute, the Buccaneer required significant braking on landing, especially at high all-up weights, and the brakes' heat would inevitably find its way into the wheel rims and tyres. When flying three sorties a day in these high ambient temperatures, it was necessary to cool the wheel units between flights and shield the wheels from the sun. In addition, I would generally opt to do carrier-type landings directly into the dummy-deck arrestor-cable when it was available, so as to reduce the landing roll and taxiing distance. We also measured tyre temperature immediately before take-off to be confident that the take-off roll itself would not push the tyres above critical limits during flight. Exceeding this limit risked a catastrophic explosion if the tyre's integrity reached its failure point.

Right: 1. Wings Presentation by HRH The Princess Marina at RAF Leeming, 1963. (RAF Leeming)

Below: 2. Aden from the air in 1965. RAF Khormaksar airfield can be seen in the foreground. (RAF Khormaksar)

3. The wreckage of Hawker Hunter XE623 which crashed following engine failure, 1964. (RAF Khormaksar)

4. With Captain Jonny Rose and Aden Airways DC3_VR-AAZ in the Hadramaut in 1965. (Author)

5. The CFS Skylarks Aerobatic Team at RAF Little Rissington in 1969. Left to right are Steve Holding, Bill Hobson, Self, and John Snell. Dick Snell is absent. (CFS)

6. A line up of the United States Naval Test Pilot School's fleet inventory in 1970. (USNTPS)

7. An A4B Skyhawk at the United States Naval Test Pilot School in April 1970. (Author)

8. US Army OV1 Mohawk. (USA)

9. An F4J Phantom and F8K Crusader pictured at NAS Patuxent River, 1970. (Author)

10. At the US Navy League Award ceremony in October 1970. (USNTPS)

Above: 11. 'A' Squadron's Harrier XV281 in 1970. (A&AEE)

Left: 12. Concorde icing trials underway with Canberra icing tanker, WV787, in April 1973. (A&AEE)

Right: 13. Self on arrival at MCAS Yuma, Arizona, with Buccaneer XV350, July 1973. (MCAS Yuma)

Below: 14. With Stan Atkins (left) and Peter Wilde (right) on returning from XV350 High Alpha trial. (A&AEE)

15. Buccaneer XV350 with Phantom and Skyhawk, Arizona 1973. (MCAS Yuma)

16. MRCA's first preview. Standing, from the left, are Cesare Calzoni, Fritz Soos, and self. Nils Meister is kneeling in the front. (MBB)

17. 'A' Squadron and the arrival of Tornado ADV, ZA254, April 1982. (A&AEE)

18. An untidy bomb release. (A&AEE)

Above: 19. ADV refuelling on the way to Aberporth Range for Supersonic Skyflash firing, June 1983. (A&AEE)

Left: 20. Buccaneer refuelling from Vulcan B2(K) Tanker, September 1983. (A&AEE)

21. Sea Harrier ski jump, Yeovilton, May 1982. (A&AEE)

22. Harrier GR3 742 during the first in-flight refuelling from a Hercules, June 1982. (A&AEE)

Above: 23. Tornado ADV ZA254 during High Alpha handling trials, March 1983. (A&AEE)

Left: 24. Scout XT647 crash at Boscombe Down, 28 September 1982. (A&AEE)

Right: 25. Last flight as OC 'A' Squadron, September 1983. (A&AEE)

Below: 26. ETPS visit to USNTPS, 1987 – last ever fast-jet flight in the F18 Hornet. Front row, third from left: John Bolton (OC ETPS), Cdr Carlton (CO USNTPS), and self. (USNTPS)

Left: 27. Self with Colin Wilcock and Lightning T4 XS422, August 1987. (A&AEE)

Below: 28. ETPS Jaguar XX915 pictured over Salisbury Cathedral. (A&AEE)

29. Spinning the Hawk, 1986. (Sunday Telegraph)

30. A view of the Buccaneer front cockpit. (MoD)

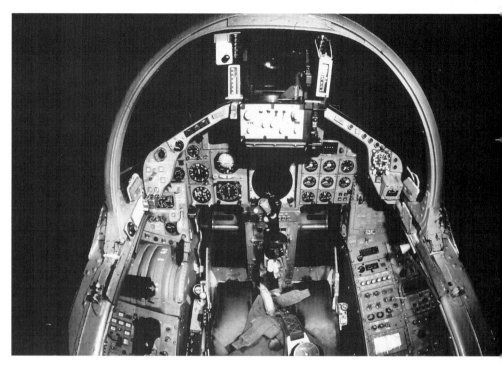

31. The front cockpit of a Tornado. (BAe)

32. With Comodoro Sapolski and the Pucara at Cordoba, Argentina, in 1991. (FAA)

Chapter 11

Testing at High Weights and Temperature

Testing the Buccaneer at high temperatures and at high all-up weight was aimed at extending the aircraft's flight-envelope. Being a former Royal Navy maritime-attack aircraft designed for employment at sea, the Buccaneer had not been tested with the RAF's heavier overland interdiction bomb loads at high ambient air temperatures. High air temperature not only reduces engine thrust (and thus aircraft performance), but it also reduces pressures in the boundary-layer blow system that enhances the Buccaneer's lift and controllability (Reynold's Number effects also play a part in the latter – for those who may know about such things). Our trials were therefore designed to investigate the effects of high ambient air temperature and high all-up weight on single-engine performance and on single-engine, low-speed (high angle-of-attack) handling.

If you suffer an engine failure before take-off or immediately after unstick (i.e. before the undercarriage is retracted), you would elect to stay on the ground or put the aircraft back down onto it PDQ (you'd probably have no choice anyway). The Buccaneer, like many modern fast-jets of the era, was equipped with a tail hook, which in conjunction with a runway arrestor cable (as previously described), could stop the aircraft before it ran out of tarmac, even if the decision to abort was taken quite late in the take-off run. We did not need to test the aborted take-off case, however, since the arrest procedure was already well documented and understood. Our tests instead were to establish the minimum airspeeds ('safe' speeds) at which an engine failure could be accommodated after take off, that is once the undercarriage had been retracted, and with no chance of putting the aircraft back down onto the runway. Once the undercarriage was on its way up, by the way, there was never enough runway (or time) left to land back onto it (and stop safely), even with Yuma's 13,000ft of tarmac.

To assess these single-engine failure cases, almost all of my thirty-seven take-offs at Yuma involved chopping an engine soon after getting airborne to establish where the safe limits lay. Our tests were designed to find the relevant 'safe' speeds across a representative range of high all-up weights and high outside-air-temperatures, and were carried out at incrementally lower airspeeds until we came upon the speed limit below which it was unsafe to go further. For each test point, the technique would be to accelerate to the test airspeed, and then sharply chop one engine to idle power, thereby simulating an engine failure (leaving the other at full power). I would then sit back and wait to see what happened next.

We defined 'safe' speed as the minimum speed at which the aircraft could be safely flown away after an engine failure. If after cutting the engine, positive acceleration could be maintained and/or an increase in height could be achieved without losing airspeed and without reaching an excessive angle of attack, then this would be deemed 'safe'. On subsequent take-offs, I would chop an engine at progressively lower airspeeds (all other parameters being taken into account in the calculation) until I reached (or came within a gnat's whisker of) the airspeed where I could neither accelerate nor climb. Now, by adding a knot or two to this speed as a safety margin for general operational use and accounting for the idle thrust of the 'dead' engine, we would establish the 'safe' speed for the conditions of the test (temperature, weight etc). Below 'safe' speed, flight could not be sustained unless weight could be jettisoned, for example, by the release of external stores. Unless this could be done immediately, the aircraft would decelerate in level flight because of increasing drag, and wing-stall and loss of control would eventually ensue. If you found yourself in this situation and couldn't jettison weight, you would have no alternative but to resort to a Martin-Baker let down. There was, of course, an airspeed just a smidgen below the established 'safe' speed, depending upon how carefully you handled the controls, where it was just possible to maintain level flight with the airspeed neither increasing nor decreasing. In a real single-engine failure situation as critical as this (and with no stores to jettison to alleviate it), you would simply have to bide your time and wait for fuel to burn off enough to reduce weight enough to allow you to climb or accelerate – and you'd be hoping all this while that no high ground lay ahead!

In establishing these limiting conditions, it was inevitable that I would sometimes find myself in that rather uncomfortable zone which lay just outside the safe part of the flying envelope, which, without remedial reaction on my part, would have deteriorated rapidly to loss of control or an unscheduled wheels-up landing. And with perhaps only 100ft of airspace between me and the ground, I had no time to dither. Fortunately, it was fairly obvious when I had strayed into this no-go zone – that nasty sinking feeling that one gets soon after an engine cut – which prompted me to open up the 'dead' engine pretty quickly.

To ensure the consistency of our results, it was important that all simulated engine failures were carried out using exactly the same technique and flight profile. Clearly, any inadvertent changes of height during these procedures could affect our conclusions, since height can be traded off for airspeed to some extent and thus confuse the results. All the flight paths of these engine-failure profiles, therefore, were tracked by on-board instrumentation and external cameras so that any changes of profile could be compensated for in our calculations. Similarly, our results would be standardised atmospherically, and interpolated and extrapolated to build up a carpet-graph covering all the possible weight/temperature conditions for inclusion in operational flying manuals.

With the take-off test completed at the beginning of each flight, the remainder of the sortie would largely be focused on high angle-of-attack (AoA) handling assessments. For these, I flew out into the empty skies and barren scrubland of the Arizona desert to allow us to concentrate on our work without having to worry too much about conflicting air traffic or making too much noise. The objective of this work was to establish the maximum safe AoA that could be flown in single-engine flight. This entailed a seemingly endless series of 'partial' climbs with one engine at full power (the other at idle), and slowly and incrementally increasing angle-of-attack until an incipient loss of control was felt. 'Loss of control' in this context would mean just that. As AoA increased above the normal approach limit (twenty units), the aircraft would become progressively less and less responsive to control inputs, until some autonomous wandering in pitch, roll, or yaw began to take place. At this point, the aircraft would feel as if it were on a knife edge and require some very sensitive handling indeed. Any harsh use of the controls here, or any further increase in AoA, even just a small amount, would induce a departure from controlled flight (and in the Buccaneer, this would usually be irrecoverable).

A 'departure' would manifest itself as a severe wing-drop, pitch-up, pitch-down, or an auto-rotation (flick) into the first stage of a spin. In manoeuvring flight, at higher airspeed and with higher levels of kinetic energy, for example during combat manoeuvres, any or several of these manifestations could couple-up with the aircraft's inertia to create a dynamic and very violent departure. Such high-energy departures might include tumbling end-over-end or cart-wheeling, which could be so vicious as to cause the aircraft structure to break up – which is not very nice and something to avoid if you possibly can! At height, there may be enough time to react to an incipient departure and restore the aircraft to normal flight conditions before auto-rotation takes over, but at lower altitude there may not be enough room to correct the situation before the ground gets in the way. Close to the ground, any loss of control would probably also result in the loss of the aircraft – and the crew too if they were not quick enough to eject. Our single-engine 'partial' climb tests nevertheless needed to be carried out at low altitude in order to maintain the highest possible ambient air temperature, even though any actual loss of control would leave us little room for recovery or escape. Anyone observing us from the scrubland below would have wondered what the hell this massive aircraft was doing.

As already mentioned, the Buccaneer made extensive use of boundary layer control (BLC) to increase lift. Ducted high-pressure air from the engine was 'blown' across the wing leading-edge and the control surfaces and flaps to re-energise the boundary layer, increase lift, and delay the point of stall (defined as the sudden loss of lift caused by the airflow breaking away from the lifting surfaces at high AoA). This 'blow' across the wing and flaps, in particular, reduced take-off and landing airspeeds significantly, which made it possible for such a heavy aircraft to operate from relatively short British aircraft carriers such as HMS *Ark Royal* and HMS *Eagle*. With BLC, high angles of attack could safely be achieved (as long as engine RPM, and thus the 'blow' air pressure, was kept high). Up to twenty-eight units AoA with momentary excursions to thirty units could safely be maintained with careful handling. This meant that I was flying at very low airspeeds and at very high nose-up attitudes, an uncomfortable place to be in a big, heavy aircraft like the Buccaneer, especially so close to the ground. And since the Buccaneer carried most of its weapons and stores externally, each different weapon configuration presented a different aerodynamic shape and centre-of-gravity, which could also present different handling characteristics, especially near to the stall. All these likely configurations, therefore, had to be tested across a range of all-up weights in order to collect sufficient data to establish safe handling limits for all possible conditions. With all this to do, I became quite adept at high AoA 'partial' climbs, carrying out scores of them over dozens of sorties. Peter Wilde, sitting in the back cockpit taking notes for me as I provided a running commentary, was always very calm and professional, although I suspect that his left hand must never have been far away from the ejection seat handle! It was probably worse for him than for me, because I at least could feel how close we were to losing control from the feedback I was getting through my hands and feet from the stick and rudder pedals.

Other aspects of these hot weather trials included measuring the temperature and performance of aircraft systems, such as fuel, hydraulics, and electronic systems (which usually suffer in extreme heat) as well as the ability of the aircraft's air conditioning system to provide an acceptable working environment for the crew. Peter and I were wired up with monitors that measured body core temperature and heart-rate, and were subjected to a medical examination after each sortie where perspiration losses were also measured. These sorties were quite hard work in high temperature-stress situations, representative of conditions likely to be experienced by RAF operational aircrew, for example in the Middle East, and so it was important to establish if the cockpit environment could sustain such operations. I have already mentioned that rising tyre temperature could limit the number of sorties performed in these conditions. Other temperature-related effects on crew or aircraft might also prove just as limiting when a new aircraft enters service, and it was Boscombe Down's role to check all this before the aircraft was subjected to such demanding conditions operationally.

A Second Tour

I had been promoted to the rank of Squadron Leader shortly before our departure for Arizona, and on return to the UK, I was asked if I would like to remain at Boscombe Down for a second three-year tour at my new level. It did not take me long to accept. By this time, the MRCA programme was looking more certain to progress to its flight-testing phase, and I was in poll position to be the first UK military pilot to fly it. I had spent three years by then working on its cockpit and weapons systems design and so knew the aircraft intimately, even though it could not yet stand on its own undercarriage. Moreover, changes were now planned at Boscombe Down that held some interest for me. All fast-jet flight testing (and me with it, if I stayed on) was about to be consolidated into 'A' Squadron when Clive Rustin was scheduled to take over as its new CO from Mike Adams in the early summer of 1974. I found this idea appealing given my background, especially when it was proposed that I should form a small and select unit that would operate all the former 'B' Squadron fast-jets until that amalgamation took place. We called ourselves the MRCA-Buccaneer Flight and took up residence in the old 'C' (Royal Naval) Squadron accommodation, which was quite appropriate for us since we also continued to operate a fleet of Sea Vixens and Buccaneers in RN colours. In forming my select little unit, I was joined by two other fast-jet test pilots, John Blaha, a USAF exchange officer later to become an International Space Station astronaut, and Roger Beazley, a former Lightning and Phantom air-defence pilot. Two former Buccaneer navigators, Bob Kemp and Graham Swift joined us. These core aviators were supplemented from time to time by specialist navigators of 22 Joint Services Trial Unit when Martel Missile trials flights were to be performed on Sea Vixen or Buccaneer. The unit would exist for less than a year, before we were absorbed into 'A' Squadron, when I became the Fighter Test Squadron's senior test pilot with John Blaha as my deputy.

Martel Trials

Meanwhile, Buccaneer flight testing had progressed to evaluating the handling and performance of the aircraft, carrying up to four of the new Martel television-guided and anti-radar cruise missiles, which would lead on eventually to actual firing trials once the missiles' systems had been proved. These missiles, carried on the Buccaneer wing-stations, were large and heavy, and therefore had a substantial impact on handling and performance (requiring lots more of the dreaded high angle of attack work to be carried out).

In parallel with all this handling work, testing of the missile's control systems continued on our fleet of Sea Vixens, and this testing was particularly demanding for us trials pilots. The twin-boom aircraft were fitted out either as missiles or as launch aircraft, the former with a forward-looking TV camera of the type carried in the missile, the latter with a TV receiver and monitor to replicate launch-aircraft equipment. Both aircraft also carried a data-link pod with which their systems could communicate with each other. In a live firing, TV pictures would be transmitted from the missile to the launching aircraft showing the missile's view ahead, and steering instructions would then be sent back to the missile to guide it.

To test the proper functioning of this data link and guidance control process in a simulated firing, the two Sea Vixens would first settle in close-echelon formation at operationally representative launch conditions (usually within the Lyme Bay or Cardigan Bay danger areas under radar control). In a simulated firing, the 'launch' aircraft would abruptly turn away at the 'launch' point while the 'missile' aircraft would continue onwards in the direction of the target. The aerial geometry required to establish and maintain the data link between the 'missile' and 'launch' aircraft as they separated had been carefully worked out by our mathematicians. This manoeuvre was to prove critical and needed to be flown accurately. While maintaining airspeed and height, the 'launch' aircraft would turn away sharply at a precise angle of bank for the first part of the turn, and then reduce bank-angle in discrete steps until the aircraft and missile were on roughly reciprocal headings. This was not easy to perform, especially since a moment's inattention at high bank angles risked your wing-tip coming into contact with the wave tops that flashed distractingly past your left ear. If you flew this manoeuvre accurately, the data link would be established and the observer would receive the picture transmitted from the missile's camera to his monitor. He would then scan his screen for the designated target (usually a practice target moored in the Bay or a RN ship on exercise), and send steering instructions to his missile accordingly. These were simulated launches to prove the system, and the 'missile' aircraft would therefore simply overfly the target at a safe height. On subsequent live firing trials against retired warships moored as targets within a remote range safety area, the Martel proved a dramatically effective weapon.

As I reflect on all this weaponeering work through lenses coloured by conflicts of more recent times, I realise there will be some readers who find this whole subject distasteful. To those who do, perhaps I could repeat what I said in my introduction: that in the days of the Cold War, the threat of attack by forces of the USSR and Warsaw Pact demanded a vigilant posture and credible retaliatory capability. For all of us who worked within the armed forces and the Ministry of Defence – military and civilian alike – this was what drove our thinking and directed our work in whatever roles or arm we served. We strove constantly to improve our technology and capabilities to counter the changing threats that were genuinely feared, and to deter aggression against us by demonstrating our ability to retaliate meaningfully. Things were certainly simpler in those days than in more recent times, and our objectives then were clearer and more understandable to the population at large than in the remote wars of recent decades, and with that came the sense that we really were protecting our shores and guarding our freedoms. We felt proud then to be part of that, and are proud now to have done our bit to preserve the peace of those times. I don't suppose it feels any different for those involved today, although the front line of the UK's defences may sometimes be further away and more difficult to see …

Forgive me for banging on about this; as you may detect, it is a theme close to my heart!

Sea Vixen and Javelin

Now, where was I? Ah yes … I was talking about flying the Sea Vixen, another aircraft for which, like many others, I developed an affection. For its time, the Vixen was a powerful and manoeuvrable fighter and a delight to fly, although I pity the observer who sat buried in the fuselage with no useful outside view. Sea Vixen deck landings, however, posed particular difficulties that are reminiscent of my Vampire short-landing incident and of the ramp strikes of the F8 Crusader described earlier (both caused by slow engine response times). Like the Vampire (also a De Havilland twin tail-boom design), the sleek Vixen had relatively low profile drag, which meant that at approach speed, even with flaps fully extended, engine RPM was at the low end of the scale. Due to the slow response of the engine to pilot's throttle adjustments at these low engine speeds, accurate airspeed control was demanding, and the recovery from any high sink-rate that developed was sluggish. Landing on a runway, where the touch-down point is not so critical (say, plus or minus a hundred yards), pilots would generally not have thought this characteristic much of a problem. But when landing on a carrier, especially landing on a short British carrier where absolute precision was required, such a characteristic sometimes proved fatal. On the Buccaneer a similar problem was anticipated and recognised in the aircraft's design by incorporating huge petal-shaped air-brakes in a bifurcated tail cone, which would normally be fully opened on approach, creating a huge amount of extra drag. This required much higher engine RPMs at approach airspeed than would otherwise have been necessary, and thus quickened engine response. Importantly, it also kept the bleed-air pressure up so that good blow pressures were maintained for the boundary layer control system.

Before moving on from the Sea Vixen, however, I cannot resist a comparison with the delta-winged Gloster Javelin, since both these aircraft arose from similar RAF and RN requirements issued in the late 1940s and should therefore have been similar in handling and performance. The RAF needed a new aircraft to replace the Meteor for the air defence of the UK; the RN needed a new aircraft to replace the Sea Venom for the air-defence of the Fleet. In responding to these similar requirements, two aircraft manufacturers, Gloster and De Havilland came up with competing designs, the Javelin and the DH110 (Sea Vixen prototype) respectively. Interestingly, at the time it was thought a hindrance for too much daylight to penetrate the radar operator's cockpit since it might interfere with his view of the radar screen. Hence, the Vixen's observer was buried in the bowels of the fuselage, and the Javelin's navigator shrouded under an opaque rear canopy – until changed later for a transparent one when they realised it was quite useful for the navigator to see where he was going! Having flown both aircraft types, I never really understood why the RAF chose the Javelin over the Vixen, other than perhaps because of a government desire to keep both aircraft factories employed. (Perhaps there is a reader who might put me right on this?) Putting aside the observer's unhappy seating position, the Vixen felt much more agile as a fighter and would, in my opinion, have fared much better in a manoeuvring head-to-head than the Javelin (the more so if the guns had not been removed in production). The Vixen's specification was also written for aircraft-carrier operations where space aboard was limited; it was useful inches smaller in all dimensions than the Javelin (particularly in height) and could also fold its wings. The Javelin felt heavier in manoeuvre by comparison and needed to be handled with some care. Due to its high tailplane losing effectiveness in the turbulent wake of the aircraft's delta wing at high angle-of-attack, it was too easy to enter a deep stall from which recovery was difficult. It was, moreover, the only reheated aircraft that I ever flew that decelerated when reheat was engaged (below about 10,000ft altitude). Reheat was a later modification for the type, which added about twelve per cent extra thrust to offset the higher drag of the Firestreak missile configuration, and was only intended for use in high-altitude interceptions. At low altitude, the fuel-flow required by the reheat exceeded fuel pump output; on selection, the jet-pipe nozzle area increased to accommodate the anticipated higher mass-flow, but this did not materialise (because of inadequate fuel-pump output). The net result was lower thrust, deceleration – and higher fuel flow. To be fair, both aircraft were designed to be missile-launching high-altitude interceptors rather than close-combat dog-fighters, and versatility of role was not in the thinking of the day. The specified primary role of these aircraft and their 1960s and '70s successors, the Lightning and Tornado F2/3, was to shoot down hostile Soviet bombers over the North Sea with their missiles before enemy weapons could be launched.

Chapter 12

Jaguar & Cold-Weather Trials

After taking over as the new Senior Test Pilot at 'A' Squadron in the summer of 1974, I joined the Jaguar flight test team, which was then engaged in navigation and weapon system trials carrying the acronym WASAMS – weapon aiming and systems' accuracy measurement trials. Jaguar was equipped with the most modern weapon system that I had flown to date, fitted as it was with a head-up display and an inertial navigation system (INS), the latter driving a projected map display, which meant that for the first time in my single-seat flying experience I didn't need a map and a stop watch to keep track of where I was! I should explain to the technically minded, that inertial navigation systems use gyroscopes and accelerometers to establish direction, speed, and displacement using a mathematical routine. This involved single and double integration of the accelerations measured fore-and-aft, side-to-side, and vertically to give horizontal and vertical velocities and direction and distance from a known starting position. Once you told the system precisely where it was before you left dispersal (to three places of decimals), it could keep track of where you went thereafter. Such systems preceded the global positioning systems of today, and although they were not as accurate, they could not be jammed or interfered with, unlike GPS which can. Inertial platforms (with more accurate and faster-erecting solid-state gyroscopes) are still used today in combination with other navigation systems to stabilise errors from external sources and provide navigation information when GPS data is not available. Our trials were aimed at improving Jaguar weapon delivery accuracy by refining another mathematical technique, kinematic ranging, which calculated ground-target range (and thus the weapon release point) from the angular rotation rate of a target as it was approached, using the aircraft's measured height (from radar altitude) and ground-speed (from the INS).

CHAPTER 12

It was because of my newly acquired Jaguar experience (of which you didn't need much as a test pilot to be considered proficient) that I found myself asked at short notice to fly out to the Canadian Defence Forces Flight Test Center at Cold Lake in Alberta in January 1975 to complete the aircraft's cold-weather trials (on XX720). This was an unexpected opportunity that arose after Bob Cole, the original test pilot designated for the trials broke his arm playing a rather energetic version of snooker in the officers' club! (I'm told the game was called Crud.) Uncomfortable for him as it undoubtedly was, his misfortune happily provided me with an interesting outing to northern Canada (for which I am grateful). He quickly recovered, by the way, and is still flying today (2020).

Unlike hot-weather trials where testing concentrated on aircraft handling and performance (as described earlier for the Buccaneer trials in Yuma), cold-weather trials tended to concentrate on the aircraft, engine, and electronic systems' ability to cope with extremely low temperatures. The sort of low temperature failures sometimes experienced in aircraft might include, for example: rubber seals becoming brittle and thus losing their sealing properties; hydraulic fluids becoming more viscous and thus not transmitting the required pressure; fuel flow misbehaviour; canopy transparency icing or misting; and electronic systems or instruments failing due to internal freezing or moisture condensation. After the aircraft had been left out in the cold for a while to reach the required test core temperature, my job was therefore simply to get into the aircraft, start it up and fly it. If everything worked long enough for me to get back in one piece without anything serious going wrong, then the test would be considered successful and we could pass on to the next temperature test condition.

All this cold-weather testing was not especially demanding, but it meant a lot of flying over the icy wildernesses of northern Canada, keeping my fingers crossed that the aircraft systems would not let me down. I often wondered how I might survive if I had had to eject in those frigid conditions, and concluded that it would not be for long. I always carried an emergency locator beacon in my flying suit, and made sure that my intended route and timing was known to air traffic control, just in case rescue should be called for. Our tests were planned according to the forecasts of local weather conditions, which were frustratingly variable. We had to wait three weeks before we reached the minus 40°C test point that we were looking for. This needed to be the temperature at the aircraft's core, not merely at its surface, which meant that the aircraft had to be left out in the open to cold-soak, sometimes for several days and nights at a time. With a lot of waiting around for the right temperatures to be achieved, I had plenty of time to fly some of the other aircraft at the base (which sensibly were kept in a heated hangar). In fact, I flew more sorties on Canadian aircraft than I did on my own Jaguar, and these included the F104 Starfighter, the CF5 Tallon, and the T33 trainer with my Canadian test-pilot colleagues at the base.

Realising that I might need some practice to comply with the Canadian air traffic control procedures that I would encounter on my return flight to the UK, I managed to talk my way into some cross-country flying too. The most interesting of these was a trans-Canada route to Ottowa and back to carry out an acceptance air test on a CTF5 two-seater Tallon aircraft at the assembly plant. The outbound route via Winnipeg and Thunder Bay was uneventful, but on the return trip, we ran into a huge, continental-sized electric storm cell that forced us to divert deep into the USA to avoid it, so deep in fact that we finished up spending the night near Kansas City, at the Richards-Gebaur US air-force base. Making an unscheduled landing in the USA like this for Canadian pilots was covered by mutual agreements and was apparently a routine occurrence; but foreigners like me would normally have needed prior security clearance and probably a visa too. As we taxied in after landing, my Canadian colleague therefore suggested that I should pretend that I was Canadian, no doubt wanting to avoid any questioning or paperwork that might delay our departure the following day. 'After all,' he assured me over the intercom, 'you're wearing Canadian cold-weather flying clothing; you're arriving in a Canadian aircraft, and you speak roughly the same language.'

It seemed an interesting challenge. If I could put a bit of a twang into my voice, I reckoned I could pass as a north American to an unsuspecting observer, and so I agreed. After all, I had spent a year with the US Navy in Maryland and six weeks with the US Marine Corps in Yuma as well as the last four weeks at Cold Lake, and I knew the form. But it wasn't quite as easy as I thought to maintain the pretence throughout a long evening at the 'O' Club bar, swapping war stories with our friendly American hosts over glasses of ice-cold beer. And I was soon aware of their sideways glances and the questions forming in their minds as my clipped English consonants began to give me away. I had to think quickly on my feet, and told them that I came from British Columbia. Surprisingly this seemed to satisfy them, for we took off on schedule the following day.

Back at Cold Lake and still having time to spare between Jaguar flights, I also flew with Alan Pengelly, a fellow A&AEE RAF test pilot (rotary wing), who happened to be at Cold Lake at the same time, carrying out cold-weather engineering trials on the Gazelle helicopter. In addition to engineering tests similar to those that my team was conducting on Jaguar, his programme included an examination of low temperature effects on rotor blade performance. In particular, he and his team were looking at the phenomenon of rotor blade tip-stalling due to shock-wave formation at high rotor speeds and blade angles, such as might be experienced in high-rate turns or when flaring to the hover from high-rate descents or high-speed approaches. The speed of sound at minus 40°C is about 100mph lower than at plus 30°C, and thus shock-wave formation in very cold air occurs at much lower airspeeds than in temperate conditions. The concern was that the fast-spinning rotor tips were more likely to stall due to shock-wave effects at these low temperatures, and that this could lead to loss of lift and control, especially in the flare as the blades' angle-of-attack was increased.

To test for this phenomenon, Alan's programme included a series of high-rate turns and fast-decelerating flares, including auto-rotative (engine-off) descents, and I accompanied him on three of these sorties as his observer. The trick during an auto-rotative descent is to keep the blades spinning fast by descending at a fairly high rate, and to achieve this you need to push the nose down quite sharply following an engine failure (in single-engine helicopters, at least). It feels a bit like that first death-defying dive of a roller-coaster car at the top of its steep ascent when it plunges seemingly into free-fall and your stomach comes into your mouth. In the helicopter's auto-rotative descent, the blades are kept spinning much like a falling sycamore seed by the upward flow of air through the rotor. At the bottom of the descent, if the rotor speed has been maintained, the blade angle (and thus the lift) can be increased for just long enough to cushion the landing. However, if lift is reduced due to tip stall while performing this manoeuvre, the rate of descent may not be arrested sufficiently or control may be lost – with obvious and predictable consequences. I had flown helicopters on my test pilot school course and thus knew what I was in for, but these manoeuvres were at the edge of the envelope and just a little bit more challenging.

Not to be a nuisance to other users of the landing pattern at the airfield, Alan conducted his testing out of the way over a frozen lake a little distance to the north of the base. And this is where he and his regular observer crashed just three days after I had departed in my Jaguar on my journey home. Both men were injured but not seriously in the accident, and so were able immediately to operate their search-and-rescue beacon to raise the alarm. But the signal was not received at Cold Lake because the beacon's battery power had become seriously depleted by the extreme cold. It was only after the planned sortie time had elapsed and overdue action was taken, that the rescue helicopter was scrambled. Even so, it was a good thing that Alan's test location had been notified, for the beacon's weakened signal was not detected until the rescue pilot was within visual range – another example of the need to test and certify new equipment, especially for use in such extreme environments.

After four weeks at Cold Lake, we had completed our work and had given the Jaguar our seal of approval for cold-weather operations. In preparing for my return flight to the UK, we refitted the long-range ferry fuel tank to the aircraft's centreline station to supplement the two tanks normally carried on the wing pylons. This added just over 1,100 litres to the fuel load, which in theory extended the aircraft's range significantly. However, there was not much trustworthy data on fuel consumption in this configuration and more than a little doubt as to whether carrying the tank was actually worth it. Appreciating that the extra drag and weight of the tank would certainly negate some of the benefit, the question was by how much? To give me peace of mind for the long flights ahead, I carried out some long-range routes at cruise altitude and speed to measure fuel consumption. While I expected there to be some losses due to the extra drag and weight, I was astonished to discover by how much. Nearly half the fuel held in that ferry tank would be consumed simply in order to carry its weight and and extra aerodynamic drag – fuel that would not increase the aircraft's range at all. Nevertheless, while very inefficient, the other half-load of fuel was still worth having. The aircraft had a very long sea-leg across the Atlantic from St John's to Lajez in the Azores in front of it, and the extra couple of hundred miles range provided by that fuel might come in handy. With the prevailing tail-winds expected on the Eastbound crossing, I calculated that the aircraft had the range to make it without air-to-air refuelling, and was happy to have a go. Wiser heads thought better, however, and a Victor tanker was arranged to be on hand just in case. As it was to turn out, no in-flight top-up would actually be required, just as I had calculated, but it was a sensible precaution nevertheless; any number of unforeseen events might have made it necessary.

As I walked out to XX720 on the day of my departure from Cold Lake, my mukluks crunched in the snow and my breath steamed in the sub-zero air. I well remember the sorry sight that met me as my Jaguar came into view from behind a snow bank. The aircraft looked like some abandoned arctic creature standing frozen to its core, neglected and forlorn, the flag of its pitot tube cover hanging like an icicle from its nose. I like to think that it was as pleased as I was to be going home to warmer climes! It took a while to warm it up and coax it into life, but everything got going in the end, and so off I set.

MRCA First Flights

Back in the UK by mid-February (1975), I resumed handling and weapons systems test-flying, mainly on Jaguar but also on Buccaneer and Phantom. Meanwhile, MRCA prototypes P-01 and P-02 had had their first flights at the Messerschmitt Bölkow-Blohm (MBB) and British Aerospace (BAe) test centres in Germany and the UK, flown by company test pilots, Nils Meister and Paul Millett. Still working with my UK, German, and Italian military colleagues, I continued to attend MRCA cockpit and systems development meetings, and also carried out assessments with them on the cockpit mock-ups in various locations. In particular, I remember lengthy sessions in the Warton mock-up involving step-by-step 'walk-through' simulations to refine the operational sequences so as to make aircrew operating errors less likely, and to improve the presentation of electronic flight information on the front and rear cockpit displays. We spent a lot of time on cockpit ergonomics and on front/rear seat interaction, communication, and work-load. We wanted to make sure that we had got workload balanced between the cockpits and the switchery sequences logical and intuitive, that the symbology and displays were clear and unambiguous, and that the required procedures were reliably achievable even under the high stresses expected during the many different attack profiles possible. Too often, previous aircraft types had entered service with serious glitches that had been costly to put right.

Two 'Hack' Buccaneers at Warton had been fitted with suites of MRCA avionic systems and were already flying, subjecting early models of MRCA/ Tornado equipment to rigorous testing in the air. Good as the Buccaneer was in riding the turbulence at high speed and low-level, the Tornado ride at high angles of wing-sweep was much more comfortable, even above 700 knots – as I would later find out. If avionics equipment could perform satisfactorily in the Buccaneer, the kit would certainly be happy in the aircraft that it was destined for.

Our first official evaluation was scheduled to take place in the autumn of 1975. This would be the first time that we government officials would evaluate the aircraft to assure ourselves (and our respective Ministries of

Defence) that the MRCA would meet the military specifications required. This first 'Official Preview' would focus on handling and performance, since the avionic systems and displays would not be ready for evaluation for another year. As the first preview date neared, Fritz Soos, Cesare Calzoni, and I, with our respective flight test engineers and deputies, commenced formal technical training at the MBB flight test centre at Manching, near Ingolstadt in Bavaria. We moved there as a team in October and took up residence in a suite of offices within the German Flight-Test Centre (Erprobungsstelle-61 or E-61 for short). Our plan was to complete twelve official evaluation flights over the following four weeks.

Unfortunately, P-01's frequent unserviceability and the notorious Danube mist conspired to frustrate us, and in the twelve long weeks that we spent there waiting for opportunities to fly, we managed just four official flights on the prototype. In preparation for our first flights, however, each of us flew other aircraft at Manching, such as the F-104 Starfighter and the Fiat G-91, to keep our hands in and our brains up to speed, as well as to familiarise ourselves with the local flying area and procedures. We were also given one rear-seat flight each in MRCA P-01 (D-9591) to help familiarise us with the aircraft so that our first flights actually in command (when they eventually came) would not come as a complete shock. P-01's rear cockpit, however, was sparsely fitted out, having only a scattering of flight instruments and no flight or engine controls. We were really just along for the ride as 'talking ballast', but it was useful nevertheless to get some sense of the aircraft's idiosyncrasies, its sounds, and its modes of motion. As usual at that early stage in the aircraft's development, there were no simulators and no two-stick trainer versions available, so this was the best introduction we were going to get.

The near demise of MRCA P-01

My first rear-seat ride in P-01 was with Nils Meister on 23 October. This is a date seared into my memory for all the wrong reasons, for I was to come within seconds of my third ejection. The flight briefing for this test flight was held in a large meeting room in MBB's flight-test centre, full of specialists for each of the aircraft's systems, and also the ground technicians, instrumentation engineers, air traffic controllers, meteorological forecasters, and many interested onlookers who did not actually need to be present at the flight briefing per se. I mention all this, because it was the general hubbub of conversation and cross-talking in the room between all these different people (mostly in German) that partly contributed to the near-accident that was about to occur – with me sitting at the back as the unsuspecting rear-seat passenger. My German language ability was developing quite well by this stage of my association with the project, but it was not up to this. Unbeknown to me, Nils and the senior MBB flight test engineer agreed to add a nose-wheel lift-off test point to the schedule of tests that he was due to perform.

One of the tests required on any new aircraft is to establish how effective the controls (and control surfaces) are at low airspeed. A pilot needs to be able to control the aircraft adequately in roll, pitch, and yaw, right down to the stalling airspeed at the very least. His ability to do this depends upon the control moment or control power – the product of the lift force developed by a control surface when deflected and its distance from the aircraft's centre of gravity. If control effectiveness in pitch (the tailplane/taileron or elevator power) is insufficient at low airspeed, the pilot's ability to raise the aircraft's nose on take-off or during the landing flare may be restricted. If such is the case, higher take-off and landing airspeeds would be required to generate the nose-lifting forces required. This in turn would result in increased take-off and landing distances, with inhibiting consequences for operations on short or damaged runways. A standard test to check tailplane effectiveness is to carry out the take-off acceleration run with full back stick (full up elevator, or taileron in the MRCA's case) and note the airspeed at which the aircraft's nose begins to rise. This is the nose-wheel lift-off airspeed that Nils was intending to measure.

On the MRCA, however, because of its fly-by-wire command and stability augmentation system (CSAS), full up tailplane cannot be achieved in the normal full-CSAS mode during the take-off roll, even with full back stick. This is because the CSAS is clever enough to know that full up-tailplane at airspeeds below the stalling airspeed can get the aircraft into serious trouble. It thus sensibly restricts its deflection. Knowing this, Nils decided that he would slip the CSAS into a degraded mode that would circumvent the computer's cautionary intelligence and permit full up-tailplane to be achieved on his take-off run. His intention, once the nose-wheel lift-off airspeed had been achieved and noted (he told me later), was to lower the nose again, reselect the normal full CSAS mode, and then continue his take off in the usual way.

Unfortunately, neither he nor his flight-test engineer had appreciated that in the degraded CSAS mode selected, full tailplane travel would be commanded by a relatively small movement of the pilot's control stick fore and aft of its neutral position, an arc of movement well inside the stick's full range of travel. This meant that should the stick be in the fully back position (as Nils intended for this test point), it would require a substantial forward movement before any reduction of the tailplane angle would occur. Correspondingly, once the stick had been moved beyond the neutral position into the forward segment of its travel, full nose-down tailplane would be commanded by a relatively small amount of stick displacement. As was soon to be demonstrated, this was a dangerously high-gain control relationship. Imagine how difficult it would be to avoid over-controlling your car if the front wheels pivoted from full-right to full-left with only an inch or two of deflection of your steering wheel. I should add that the particular degraded CSAS mode selected for this test had not yet been cleared for flight. Nils' plan, however, was that full CSAS would be re-engaged well before the aircraft became airborne and so no intentional transgression was envisaged.

Unaware of Nils' intentions, I sat innocently in the back seat of P-01 as we taxied out for take off. Having been intimately associated with the MRCA project for nearly five years by then, I was looking forward to getting airborne in the first prototype (I did not then know how much of an occasion it would turn out to be). Nils lined up on the runway and pushed the throttles to full dry power to perform his engine checks, holding the aircraft stationary on his toe brakes while the F-4 Phantom chase pilot taxied his aircraft into a position closely alongside us. I could soon hear the roar of his engines as he too throttled up for his own engine checks. He had been briefed to release his brakes and chase after us once he had seen us getting airborne.

On a 3,2,1 countdown, spoken for the sake of the tracking cameras and those listening in the telemetry control room to our cockpit commentary, Nils released the brakes and pushed his throttles forward to maximum thrust in full reheat. The aircraft leaped forward under the surge of power and I felt myself pressed back into my seat as we accelerated fast down the runway. Unencumbered by external stores or fuel tanks, the aircraft was light and very sprightly. I monitored the airspeed as it increased, expecting rotation and lift off at around 120 knots, the calculated rotation airspeed for the aircraft's weight. And so when the aircraft lurched into the air at 90, I was taken a little by surprise. Ninety knots was far too low an airspeed for sustained flight, as was soon to be confirmed, but I thought at first that Nils was trying to impress me with some novel short take-off technique. After an uncomfortable moment of uncertainty, however, as the aircraft hung seemingly suspended some 20ft above the ground, the nose dropped sharply and we came crashing down onto the runway with a teeth-shuddering thump. But this was just the beginning of a roller-coaster ride, for the nose was already rising as we hit and were bounced back up into the air again, this time at a yet higher nose-up attitude than before.

Another pregnant moment ensued before the nose dropped sharply and we commenced another earthwards plunge. Once more we came crashing down onto the runway, hitting the ground so hard that I felt the undercarriage's hydraulic oleos hit their stops as we rebounded yet again into the air. This time the nose-up attitude became even more extreme. From my rear-seat viewpoint, I could see Nils in front of me twisting his head and shoulders in an agitated fashion, as if reaching for something on the left console. I had no idea what he could be trying to achieve, but while he was doing it, the aircraft's nose dropped again. I braced myself for another shuddering collision with the ground, wondering if the undercarriage could take much more of this. But on this occasion, our wheels only just made it to the runway, brushing its surface with the slightest touch, saved from heavy impact by a further upward swoop that first pressed me down into my seat then lifted me up against my shoulder straps as the aircraft bunted over into yet another nose-dive. The underlying flight path thankfully had now begun to assume a generally upwards trajectory; but superimposed upon this shallow climb was a dangerously divergent longitudinal oscillation of ever-growing proportions. The forward view through the windscreen soon became a blur, filling alternately first with a picture of the ground and then with sky. With each succeeding cycle, the nose attitude was becoming more extreme, the airflow protesting noisily as the wings strained to arrest each subsequent descent. I began to fear that the next downward plunge might be our last.

If all this was alarming to us from the cockpit, pandemonium was breaking out below. The air traffic controller, watching the aerial spectacle in horror from his tower, had triggered the crash alarm, expecting to see a conflagration on his runway pretty soon. A klaxon sounded loudly across the base and several fire and rescue vehicles, their lights flashing and sirens wailing, at once shot from their stations and followed us across the airfield in hot pursuit, anticipating no doubt that they would soon have some work to do. But this was not all. The F-4 Phantom chase pilot, having seen our first (premature) rotation and believing it to be a normal take-off, had let off his brakes as per his brief, and had come roaring up fast behind us in full reheat. The F-4 in reheat accelerates extremely quickly, and only at the last moment did its pilot realise that he was on a collision course with Nils and me ahead. Throwing his flight-controls over, he broke away sharply in a cloud of black exhaust smoke, the thunder of his engines rattling all the windows on the base. Our UK flight-test engineers, Lionel Phillips, Alan Ashton, and other members of our flight-test team sitting in the darkened telemetry room and thus blind to the chaotic scene outside, thought the snaking traces on their monitors must be some kind of fault.

But just as our aircraft pitched over into what was very likely to have been our final nose dive – and just as my hands were beginning to move towards the ejection seat handle between my legs – the oscillations stopped, and the aircraft settled into a stable, climbing attitude. It was like the sudden calm after a wild storm. Having been all over the place during our helter-skelter ride, the airspeed now began steadily to rise, the altitude slowly to increase. I heard the hydraulic whine of the undercarriage and flaps being retracted and felt the trim change as we accelerated smoothly towards climbing speed, but it was a good thirty seconds before anyone spoke, either in the cockpit or from the telemetry room.

It was Nils who broke the silence. 'Right,' he said, catching his breath, 'what's the first test point, Ron?' It was as if we had not just achieved one – and a salutary one at that, for with our narrow escape we had inadvertently identified a control problem that would need some attention. I cannot now remember precisely what the next test point on my check list was, but I read it out without further comment. Whatever it was, we climbed to a few thousand feet and set ourselves up for it as the F-4 chase aircraft came back into close formation alongside. I glanced across at its pilot, Karl Steuer, an E-61 government test-pilot colleague and friend, and raised my hands in the air as if to say 'Blimey, what the hell happened there?' He shook his head silently and shrugged his shoulders, apparently as puzzled and as taken-aback as I was. In the front cockpit of P-01, Nils was unusually quiet, probably still mulling over what had happened; and we were not far into our test schedule when he suggested that we call it a day and return to base.

In the debrief (not a literal post-mortem, thankfully), the cause of our involuntary fairground ride became clear. An examination of the instrumentation traces showed the tailplane angle moving from fully up to fully down over a relatively narrow arc of Nils' control stick movements. This high-gain tailplane authority, combined with apparent 'dead bands' at both ends of stick travel, made over-controlling almost inevitable. We could see that after raising the nose-wheel (at about 90 knots), Nils had moved his control stick forward to check the rotation by an amount that would have been about right if the CSAS had been in its Normal mode. But in the degraded mode that had been selected for this test point, his control input had not been enough. The tailplane angle had remained fully up-deflected, and as a consequence, the nose had continued to rotate upwards at a high rate until the aircraft had thrown itself inelegantly into the air. Seeing what he thought to be an un-commanded over-rotation, Nils' natural and instinctive reaction had been to push the stick forward rapidly in an attempt to check the motion. Distracted by this urgent need for corrective action, he did not immediately re-engage the Normal CSAS mode as he had intended. And so now the control system's high gearing again caught him by surprise. Even though his stick never reached the fully-forward position, the tailplane had slammed fully nose-down, causing the aircraft at once to rotate rapidly earthwards.

The MRCA's tailplane was as big as the proverbial barn-door and had enormous power, and because of the high gearing in this degraded flying mode, even moderate stick movement fore and aft had caused full-range deflections. It is almost impossible for a pilot in such dynamic circumstances to avoid getting his stick movement out of phase with the aircraft's motion and thus find himself inadvertently driving the oscillation to extremes. (This, by the way, is called a pilot induced oscillation – the notorious PIO – a topical subject to which I will be returning.)

Recognising the situation for what it was, Nils now attempted to reselect the normal CSAS mode, the selector button for which was positioned behind the throttle quadrant on the cockpit's left console. Locating this button, however, was difficult with the snatched glances that he could spare while fighting for control. And because of the wildly alternating 'g' forces that were throwing us about in our seats, he had found himself physically unable to guide his left hand accurately enough to bring his finger into the correct position. At the same time, his right hand, instructed by that part of his brain not engaged with this desperate digital search, was attempting but failing to bring the aircraft back under control. In the nick of time, his finger finally found its mark and pressed the button that it had been groping for – and hence the day, the aircraft, and most likely the two of us were saved.

Chapter 13

Meanwhile, back home in the UK, Caroline was plotting a divergent flight-path of her own. The family had by this time lived in our cottage in Shrewton for five-and-a-half years, and we sought more space. Caroline in particular had always had a hankering to live near water (and still does), and during the weeks that I had been away in Germany waiting impatiently for my chance to fly MRCA, Till Cottage came onto the market. She rang me straight away, very enthusiastic about it, and sent me some photographs with her sketches of the floor plan and the layout of the garden. The cottage was a typical two-up, two-down, former farm-labourer's hovel, built of cob under a thatched roof, to which had been added a lean-to bathroom and WC, and off-peak electric heating. Water supply and drainage were basic, waste treatment was by way of a septic tank and its water supply came directly from the cottage's own well, which at least used an electric pump to fill the header tank (rather than a pulley and pail). The cottage had only two bedrooms, making it a bit tight for a growing family of four, but Caroline thought that it could be extended without spoiling its character. And anyway, it had the very thing that she had been looking for – a beautiful location and about a hundred yards of river frontage on the River Till. The Till was a typical Wiltshire chalk stream of crystal clear water, perhaps only a foot or two deep at most but quite wide and freely-flowing. Although the cottage was small, it was very pretty, with roses around the door and a south-facing aspect – the chocolate box picture of our dreams. The spot was a delightfully peaceful haven for wildlife, and the river was full of trout. I flew home as soon as I could to view it and fell in love with it too. We decided there and then to put proceedings in hand. For the record, you could buy all this for £21,000, which was then a bit of a stretch for us, but one that we never regretted.

First MRCA Preview Flight

Leaving Caroline to to deal with the purchase of Till Cottage and the selling of our current property, I returned to Manching to fly the MRCA for myself on 28 November. This time, Nils Meister was strapped into the rear seat as my observer, and I gave him a far less exciting ride than he had given me the month before. My flight time was one hour and sixteen minutes, and as the first ever RAF pilot to fly the aircraft, I made the most of it, exploring all the available corners of the permitted envelope, with scores of different test manoeuvres and simulated operational tasks. Tests included all the usual ones designed to ferret out any likely issues or handling problems at an early stage in development so that they could be put right before entering service. These included wind-up turns over a range of airspeeds and wing-sweep settings to check stick-force per G in manoeuvring flight, and numerous forced excitations of the longitudinal short-period and lateral-directional (dutch-roll) oscillations to check damping characteristics. A range of steady-heading side-slips, full-stick deflection rolls, level accelerations and decelerations, and some high angle-of-attack test points were also evaluated. Tests were conducted in the normal and degraded CSAS modes, the latter exhibiting reduced stability and different control characteristics (as I knew quite well from my earlier experience in the back seat). Cloud conditions on the day made constant manoeuvring necessary to remain within an acceptably clear test area, which made it hard work for the chase pilot to keep up. The summary paragraph on the cover of my sixteen-page report is interesting to reread now, forty-five years later. Actually, there was not much wrong with the aircraft and therefore not much of real significance to report.

In Full CSAS mode (C5 computer), dutch-roll and longitudinal short-period damping was mostly dead-beat throughout the flown envelope, and the aircraft was generally pleasant to handle. Controls were well harmonised and stick-force per G was consistent with the aircraft's role and moderate over the whole speed range (about 5lb/G at wing-sweep settings 45° and 66° and 6lb/G to 7lb/G at 25° wing-sweep). Configuration changes, including wing-sweeps over the full range, produced only small and sometimes almost unnoticeable trim changes. The general level of airframe vibration/buffet above 300 KIAS (knots) was not excessive, but high enough to detract from the generally favourable impression. In reversionary CSAS modes, aircraft handling was generally good, but in YAW ND (yaw no damping) mode, rolling manoeuvres could not be coordinated – the aircraft felt untidy and closed-loop tracking tasks such as target tracking are likely to be imprecise. Trim changes in PITCH DL (direct-link mode) during flap selections were large, but trim-ability generally was good. Control mechanical characteristics were unobtrusive and satisfactory, but large stick oscillations were noted during rap inputs (although without perceptible aircraft reaction). Handling during circuits and landings in Pitch DL mode was good.

The airframe buffet/vibration referred to in the above summary manifested itself as a vertical bouncing motion above about 450 knots, which was disconcerting and became uncomfortable as speed increased further. The problem was caused by airflow entrainment over the rear fuselage, where the airflow met the engine exhaust in the cut-away section at the base of the fin. After further experimentation, this was corrected in later prototypes and for production aircraft by inserting a solid semi-bullet-shaped fillet into the gap.

I pronounced myself impressed with the aircraft overall, and concluded that the MRCA was likely to be pretty good for its intended low-level interdiction role (it does not do for a government test pilot to enthuse too much!). Apart from the buffet and bounce phenomenon referred to above, the aircraft was a delight to fly. Fritz Soos, my German government colleague had made his first flight two days before me, and Cesare Calzoni, from the Italian flight test centre at Pratica di Mare, made his two days after. We all thought the same. Between us, we were able to cover all the essential testing required at that early stage of development to give it the thumbs up. More work would be required on CSAS software, and the avionics and weapons-system had yet to be evaluated, but we all knew even then that we had a potential winner on our hands. The three of us had worked on the MRCA project for many years, and so these flights represented a significant and satisfactory milestone for us all. Poor weather and aircraft unserviceability continued to dog us, however, and we finally had to throw in the towel in mid-December, having completed only four flights of our planned twelve. Nevertheless, that single flight of mine in November 1975 made me the first RAF pilot to have flown the MRCA, and I remained the only RAF pilot with any MRCA front-seat experience until the second official preview in the summer of the following year.

I returned home just in time for Christmas, but as soon as the festivities were over, we were busily preparing for our house move to Till Cottage. We had completed the purchase of our new home in only eight weeks since first seeing it, which included the Christmas hiatus. Now, with snow drops and crocuses popping up all over our new riverside garden, we looked forward properly to settling in. But only a few days later, while still on leave and still unpacking boxes, I received a telephone call from John Wilkinson, the then Superintendent of Flying Division at Boscombe Down, telling me that there had been an accident involving P-05, the Italian MRCA prototype, at the Aeritalia test base at Casselle near Turin. Thankfully, no one had been injured, but a board of enquiry was to be convened the following week and the panel would need to include suitably qualified representatives from all three nations. He hesitated before going on, and I think he expected me to guess what was coming next. I packed my bags and was on my way to Italy two days later.

MRCA P-05's PIO

The P-05 accident proved to be related to the divergent oscillation incident on P-01, in which I had been an incidental participant just six weeks or so before. Pietro Trevisan, the Aeritalia chief test pilot, was returning solo for a landing following a test flight (the rear cockpit was occupied by instrumentation equipment), and he had set himself up for a long straight-in approach to Casselle's main runway. The aircraft was in full CSAS mode rather than the degraded mode previously described in P-01, thus control responses should have been perfectly normal. But Pietro reported that at a late stage in the approach, just as he was preparing for his landing flare, the nose appeared to rear up abruptly without apparently any stick input on his part. It was a calm day, and so wind-sheer or air turbulence could not have been the cause. His natural reaction, he said, was to check forward on the stick to lower the nose to its normal attitude, but in doing so, the aircraft appeared to over-react to his input, responding with a sharp nose-down rotation. Being so close to the ground – perhaps only 30ft or so above the runway at this point – Pietro had neither the time nor the airspace to arrest the ensuing descent or prevent the aircraft from hitting the ground so heavily that all three undercarriage legs collapsed. From that point onwards, there was nothing further that Pietro could do – other than hope for the best – as P-05 slid along the runway on its belly in a shower of sparks and smoke – with Casselle's fleet of fire and rescue vehicles racing along behind him. Eventually, the aircraft came to a halt, still on the tarmac but askew, leaving a long trail of metal debris behind it. The fire crew were on the scene within seconds and quickly doused the incandescent metal of P-05's buckled underside. Fortunately, no fire developed. From photographs that I later saw of the aircraft in this sorry state, it looked as if it was otherwise undamaged and would soon be back on its feet. The subsequent examination, however, showed that the entire structure had been badly distorted by the impact and would need extensive repair It would not fly again until March 1978, over two years later, after what amounted almost to a complete rebuild.

This accident on P-05, and the near-accident on P-01 were examples of the sort of teething problems that are sometimes experienced with fly-by-wire computer-controlled flight-control systems in the early days of development. In aircraft with conventional mechanical/hydraulic control systems, the relationship between the pilot's flying controls and the aerodynamic control surfaces is (more or less) fixed by mechanical linkages such as push-rods or cables. By contrast, except for failure conditions when emergency mechanical reversion is possible, a Tornado pilot's control commands are communicated entirely by way of digital signals via a flight-control computer. This 'black box' also receives air-data (such as airspeed, angle of attack, Mach number, and altitude), dynamic data (such as roll rate, pitch rate, and yaw rate), and aircraft configuration data (such as wing sweep, flap and undercarriage position). All these data are fed digitally into equations embedded within the flight-control computer's software where algorithms then determine what instructions are sent to the control-surface actuators. Signals from the pilot's controls may arrive at the computer as his commands – for example to adjust roll or pitch attitude – but the control surface responses to these commands depend on the software. In effect, the computer decides what control surface deflection is appropriate to achieve what it *thinks* the pilot wants; but this is tempered by any restrictions that the software imposes to prevent him doing anything that is considered inappropriate or dangerous. The computer could decide, for example, that the pilot is straying into an unsafe corner of the flight envelope and modify the signals sent to the control-surface actuators to prevent the excursion. It could even ignore the pilot's commands entirely, which might occur, for example, if the pilot tries to pull the nose up when the angle-of-attack is already too close to the stall angle.

Similarly, if the aircraft has an unstable tendency, the computer's sensors might detect that loss of control is imminent and signal the control surfaces to act independently of, or even counter to, the pilot's control inputs in order to dampen unwanted oscillations or divergences. The pilot is therefore in the hands of the software, and it takes time in a new aircraft for this to be developed to a state where it is entirely trustworthy and glitch-free. The formulas and algorithms in the programme will initially have been developed using mathematical modelling and simulations, but these need actual flight data to refine them. And clearly, at an early stage of a new aircraft's development, only limited actual data will have been accumulated – not enough at first to guarantee a perfect job. Software development for such systems will thus inevitably be iterative: fly – analyse – adjust the software, fly again – analyse again – adjust a bit more, and so on, until full confidence is attained through flight testing to demonstrate that the system will always behave as it should. If the formulas are wrong, even by just a little, the computer could become renegade rather than guardian – a clumsy, mischievous adolescent rather than a wise old friend upon whom you can depend to act in your best interests.

An examination of P-05's instrumentation traces following its accident revealed that Pietro was the victim of just such a problem – that the not-yet-perfected software had led him unwittingly into a pilot induced oscillation (PIO).

A PIO is not, as its name might imply, the result of pilot mishandling. The phenomenon is the result of an interaction between the aircraft's motion and the pilot's control behaviour, which conspire to create an oscillation that can become divergent and indeed catastrophic (as it nearly was in this case). Such oscillations occur only when the pilot attempts to impose his will on the aircraft – indeed, left to its own devices, the aircraft may be stable and recover of its own accord. But because the pilot's actions depend in part on the aircraft's response to his/her commands, the aircraft and pilot dynamics interact to form a closed-loop feedback-control system that can drive itself divergent. The pilot is said to be 'operating closed-loop' or to be 'in the loop'. Pilot induced oscillations are, therefore, closed-loop instabilities, which generally only occur when the pilot/aircraft system is operating at high gain (intense pilot activity), such as is typical of tasks like in-flight refuelling, tracking a target, and flying in close formation. But it can also, as in this case, occur on landing, where the pilot is attempting to touch down at a precise point on the runway, and thus is quite active with his control inputs.

Well before Pietro became aware of it, an embryo PIO was evident on the instrumentation traces. But the oscillation built quickly over a period of about ten seconds, and it only revealed itself for what it was in its penultimate cycle. This was the sudden, apparently un-commanded, rearing up of the nose that Pietro reported. He had inadvertently and unknowingly been driving the oscillation to greater and greater amplitude, a closed-loop of action and reaction between pilot and aircraft, becoming more extreme with each successive cycle. Nils, Pietro and I had all nearly come a cropper through the renegade characteristics of a computerised flight-control system for which software was still at a relatively early stage of development. Others have not been so lucky in accidents right up to the present day. The primary cause of the two Boeing 737-Max crashes in 2019, when all 300+ souls on board the two aircraft lost their lives, is suspected to have been renegade commands from the flight control augmentation system (MCAS). In this case, spurious signals from a single sensor (the angle-of-attack probe, I believe), appears to have caused the MCAS to command full nose-down tailplane trim, which the pilots were apparently unable to overcome. Simulations based on mathematical modelling are only as good as the cases evaluated and the assumptions made, and thus should never be relied upon without flight test verification of all the possible operational situations including failure modes. It would also seem a bit reckless for such powerful, safety-critical flight-control systems to depend on single sources of critical data, such as angle-of-attack. The MRCA system incorporated duplex and triplex data sources with monitoring that would out-vote anomalies. If an anomaly was detected in any one of the critical data channels, the pilot would be alerted and the system would revert automatically to a safe alternate mode.

Whatever the design, automated flight control systems should also be provided with an instinctive override to a safe reversionary mode. Simply taking hold of, or grabbing the controls (in panic) and exerting a force exceeding a suitable pre-set break-out pressure should be enough to revert. If you are wrestling with a renegade system, taking your hands off the controls to operate remote switches could waste critical time. In the 737-Max story, the danger posed by a trim run-away does not appear to have been anticipated. Consequently, nor was the need to establish, document, and train for the appropriate pilot intervention should such a failure occur. Neither was it apparently thought necessary to provide the operators with a ready and instinctive method of reversion to a safe, more direct flying mode. Lessons from our past experience of automated and computerised control systems should have taught aircraft manufacturers the error of such omissions a long time ago.

Chapter 14

The NATO MRCA Management Agency – Munich

After the P-05 accident Board of Enquiry had been concluded, I returned to Boscombe in February 1976, and was called in to see John Wilkinson again in his office. He was naturally interested in the Board of Inquiry's findings, but he also had a proposal to put to me. The NATO MRCA project management agency (NAMMA) in Munich had been in touch. They were looking for an RAF test pilot to join its flight test programme management office (SE10) in Munich, and I had been identified as a likely candidate. Would I please, therefore, fly out to Munich to attend an interview? This proposition did not take me entirely by surprise because the idea had already been mooted during those twelve long weeks that I had spent drumming my fingers in Manching waiting for the fog to clear. Volker Storch, the section leader of SE10, and Berndt Bau, his number two, both former German Government flight-test engineers at Manching and both still living in nearby Ingolstadt, had sometimes joined the social gatherings of our tri-national group, and we had got to know each other reasonably well. The outcome of my interview in Munich was therefore almost a foregone conclusion, especially since there was no other Brit who could meet the qualification criteria quite so well as I could at that time.

When I returned to Boscombe a few days later, John Wilkinson telephoned me to tell me that I had been selected for the post (from a shortlist of one), but also that I should understand that the three years I would spend in Munich would be as a NATO officer, and that the time served as such would therefore not count towards my RAF service. Were I to accept the job, therefore, any plans that I had to retire at the age of thirty-eight (the optional retirement age), would have to be delayed. I was then thirty-five years old, and already contemplating switching career paths, but three years living in Bavaria was too good an opportunity to miss. Having already completed two full test-flying tours, I was due for a staff tour anyway, probably in the Ministry of Defence in Whitehall, and so Munich most certainly represented a far more enticing prospect.

I arrived in Munich in March 1976. The family would not be joining me until September in time for the start of the new school year, and so I found myself a single bed-sit not far from the Olympic stadium and a short drive to Arabellastraße in which the NAMMA building was located. I had already spent too much time away on one duty excursion or another as previously described, and so I made my first six months more bearable with monthly long-weekend trips home. Communication was more difficult in those days. With no landline in my bed-sit, and before the time of mobile telephones, email, Whatsapp, and Facetime, keeping in touch required a search for a vacant telephone kiosk and a pocketful of Deutschmarks and pfennigs. Letters home, moreover, took a week for the round trip.

After some searching, I found a house to rent in Ottobrunn, a suburb in the south of the city and an hour's commute by S-Bahn from my office. Caroline, Sonja, Andrew, and our furniture all arrived on the same day in early September and we moved into our new home straight away. For S and A in particular, it was a baptism by fire. At 10 and 8 years old respectively, it was sink or swim for them as they entered the local Volkschule the following week with only a smattering of German; but they quickly found their feet. Caroline settled into German life quickly too, finding herself employment as part-time secretary for the MRCA avionics department in the Messerschmidt-Bölkow-Blohm works, situated just a mile or so from our home. She used to cycle to the works four days a week, and was soon able to compete with me on avionic acronyms. As for me, in those months of single living before the family had arrived, I had made a real effort to improve my imperfect German by attending the Berlitz language school three nights a week. If never coming close to Caroline's level of proficiency in the language, I could at least usually raise a laugh among our German friends – if for the wrong reasons.

I had joined section ten (SE10) of the systems engineering department, the section responsible for managing the MRCA flight test programme. With my arrival, SE10 comprised two Germans, two Brits, and one Italian in four offices on the fourth floor. Volker Storch, as section leader of SE10, had an office to himself; I shared another with Berndt Bau; and Ted Pearson and Adolfo Persichini (succeeded later by Paolo Corazziari) shared a third. Our secretary, Linda Weeks, in her own office with all the filing cabinets and the section's coffee machine, made up the sixth member of our team.

With names like these, there are no prizes for guessing my new colleagues' nationalities! We made a close-knit and happy team, earning a reputation with our external clients as 'efficient, businesslike, and firm', and internally within NAMMA as 'that noisy outfit at the end of the SE corridor who always seemed to be having far too much fun' (which we were!). Between the five of us, we managed the different elements of the flight test programme in the three nations' official and manufacturers' fight-test centres. We also ran the monthly flight-test management meetings that were attended by representatives of the three nations' air staffs, the national military clearance authorities, and the aircraft and systems manufacturers. Ted and Adolfo managed the avionics and engineering development programme, Berndt managed the performance and handling trials programme, and Volker oversaw and knitted together the whole thing. My own responsibilities included managing the armament flight-test programme, monitoring all flight test activity for flight-safety issues (of which I had become an unintentional expert), and overseeing the development of head-up display (HUD) symbology. This third part of my responsibilities – MRCA HUD symbology development – turned out to be the most interesting, and one that might therefore be worth a few lines before I move on.

Head-up displays (HUDs) are a standard fit in most military and some of the more modern civil aircraft these days (and now even in some motor vehicles too), but in the 1970s, the only HUD-equipped aircraft operated by the three MRCA nations were Jaguar and Harrier, both then flown by the RAF. The Buccaneer's strike-sight might be described as a HUD of sorts, but it was much more limited in its field of view and included no flight data as such. This was also true of the strike sights in the most modern fast-jets then operated by Germany and Italy, the F-4 Phantom and F-104 Starfighter, for example.

Stating the obvious, the innovation and advantage of head-up displays lies in the name. A head-up display presents flight and other operational information directly in front of the pilot's eyes – superimposed on his view of the outside world – thus allowing him to fly accurately while keeping his head up and his eyes out of the cockpit (where they should be). HUD information and data is presented as an array of graphic, electronically-generated symbols, which represent a miniaturised version of what is also available from the 'head-down' cockpit instruments or displays. Because this symbology is optically collimated at infinity (theoretically), it also means that the pilot's eyes do not need to refocus between reading the data and viewing the world beyond – possibly saving critical moments if that speck of dirt on the windscreen turns out to be a fast-jet coming the other way!

For those unfamiliar with the technology, the head-up display projects computer-generated flight data, navigation, and weapon-aiming information onto an angled, beam-splitting transparent glass reflector (the combiner) located inside the aircraft's front windscreen (or on to the windscreen itself if optically perfect) – right in front of the pilot's eyes. From the pilot's eye position, it looks as if this data is written in the sky in an array of alpha/numeric digital and analogue symbols (usually in an iridescent green colour), which variously represent all the information necessary to fly and operate the aircraft. The display will usually include flight and operational information such as: (bear with me, it's a long list!): flight-level, altitude, or height (above ground); airspeed and/or Mach-number; heading; vertical speed; angle-of-attack; attitude; flight director; velocity vector; the course to steer; planned track and distance-off; timing fast or slow on flight-plan; range to the next turning point or target; glide-path information for instrument landings; target and weapon-aiming markers such as laser-marking, air-to-air missile seeker-head aiming point, a bomb-fall line, and a continuously computed weapon impact point (all self-explanatory, I hope).

With such a large amount of data to present within a relatively narrow field of view, great ingenuity and care is needed (not to mention a lot of trial and error) in the design of the symbology. This is still true even for more modern HUDs that have wider fields of view because the data still needs to be presented within the pilot's focal zone without too much of a need to scan. If you have to move your eyes around a lot to interrogate the different elements of a display, precious time is wasted. This is why the helmet-mounted displays that are employed in more modern military aircraft such as the new F-35 Lightning can be so helpful, since they permit critical data to remain within the pilot's focal zone even when he swings his view from side to side.

In the early days of HUD development, the relative unsteadiness of inertial-platforms employed in Harrier and Jaguar caused jitter in the pitch-attitude bars of the display, which was disconcerting. In order to dampen or remove this jitter, these bars (the lines representing degrees of pitch above and below the horizon) were scaled at 5:1, which meant that for each 5° of actual pitch change, the attitude bars shown on the HUD would move only 1°. Possibly expecting the same inertial-platform unsteadiness (or possibly just not thinking it through), pitch-attitude symbology in the MRCA HUD had also been designed by its makers to work on the same scaling. After all, it was argued, 5:1 scaling had seemed acceptable for the Jaguar and Harrier and so it should be good enough for the MRCA, shouldn't it? But such a scaling meant that there was an apparent disconnect between the pitch attitude as displayed on the HUD and the pitch attitude as seen by the pilot by reference to the actual horizon ahead. Jaguar and Harrier pilots had seemingly adapted to this disconnect, but any objective analysis would have to conclude that the arrangement was potentially confusing and likely to be disorientating in some circumstances. I had certainly found it so.

On the other side of the Atlantic, modern US fast-jet HUDs were already employing 1:1 (real-world) pitch-attitude scaling, and I wondered, therefore, why the MRCA HUD was not following suit. The inertial platform in the MRCA was solid-state and an order of magnitude more stable and more accurate than those of the Jaguar and Harrier, I argued, and so the more realistic 1:1 scaling should now be sufficiently jitter-free to make it acceptable.

Being the only official pilot within the MRCA cockpit design group to have HUD experience, and now having some clout as a flight-test programme manager, I argued for and was given the authority to run a HUD symbology development programme. In discussion with the then 'A' Squadron MRCA project test pilot – my successor, Don Thomas – we put together a flight test programme to test 1:1 scaling and also develop and test other symbology improvements, such as the use of analogue scales for altitude and airspeed, which had until then been entirely digital. The outcome was almost predictable. All the pilots who flew the new display design and the 1:1 scaling were immediately taken with the proposed improvements, and thus the new symbology standard was adopted for production. That symbology remained substantially unchanged throughout the forty-five years of the MRCA/Tornado's operational life.

The NAMMA work-culture was very European. Where long hours would have been the expectation in the dingy corridors of the UK Ministry of Defence, even on routine, non-emergency days, the working week in our NATO organization was normally expected not to exceed 37½ hours. We worked punctually from 08.00 am until 5.00 pm Monday to Thursday with an hour for lunch, and finished mid-Friday afternoon, making every weekend feel like a long weekend. And being an international organization, we celebrated the different national holidays of all three nations represented in the project, which meant extra days off, over and above our six-week annual leave allocation. For all that, I cannot say that we were any less productive. Indeed, being able to compare the two cultures, I came to realise that the correlation between working hours and productive output was often perverse. Our working conditions were good too. Unlike the worn-out decor and dark offices of Whitehall, we worked in a modern, light and airy office block with good facilities. In those days, before personal desk-top computers took away their role, secretaries assigned to each section within NAMMA did all our typing and filing (and made the coffee too!). Until the seven-story Hypo Bank building was erected to obscure the view from my office window, I could see the twin domes of the city-centre's iconic Frauenkirche, and on a clear day the distant Zügsptize, which at around 10,000ft, was the highest mountain of the German Alps. If I had to fly a desk at all in my RAF career, this one in Arabellastrasse must have been one of the best!

Chapter 15

Unity is Strength – Inscription on the National Defence College crest.

In June 1979, after three straight years in Munich, I joined the six-month joint-service staff college course at the National Defence College at Latimer House in Buckinghamshire, one of those stately old mansions requisitioned by the Government during WWII for military use. It was, by the way, one of three stately homes in the UK where captured U-boat crews and Lufwaffe pilots were held before being transferred to prisoner of war camps. Run by an Army major-general, the NDC course was run at a gentlemanly pace, which allowed me to travel up from Wiltshire on Monday mornings and back on Friday afternoons, which meant that I got a proper week-end. The course was designed to broaden our perspectives – to meld us dark-blue, brown, and light-blue types into a fetching shade of purple and prepare us for central defence-staff posts in the MoD. To understand each other's roles better, we went on manoeuvres with the British Army on the Rhine, to sea with the Royal Navy, and came face to face with Russian and East German soldiers in East Berlin. I finished the course as a wing commander with a posting to the Defence Operational Requirements Staffs in Whitehall under a deputy chief of the defence staff (DCDS (OR)), then a Royal Navy admiral. With RN and Army colleagues, I worked on specifications for new items on the military wish lists of all three services. My particular project as DOR2(Air) was the EH101, a joint Italian/UK project that would eventually become the Merlin helicopter, but was then still in its early stages of definition to meet the requirements of the Italian and British navies. I took part in the early debate with naval staffs in deciding if this new aircraft should have two or three engines to meet their mission requirements. Although the extra engine would mean more weight and higher costs, the increased fuel capacity and range of the bigger airframe swung the argument in favour of three; its performance and survival chances in an engine-failure case were, of course, also considerably better. The aircraft would not fly until 1987 and would not enter service until ten years after that, but versions of Merlin continue in service today (2020) with armed services around the world, including its use as a presidential VIP transport in the USA and India. When I spot a Merlin flying today, I wistfully remember the tiny part that I played in its conception forty years ago.

Mercifully, I was given a reprieve after only nine months in Whitehall's corridors of power this first time round (there would be more to come later). Summoned by the admiral to his walnut-panelled office one morning in September, I could tell at once by his demeanour that he was not a happy man. At first I wondered if I had made some dreadful cock-up and was to be given a dressing down for it, for I was not invited to sit. Not hiding his pique as I stood before him apprehensively, he told me that despite his protests, I was to be withdrawn immediately from his department to be returned to test flying duties at Boscombe Down. This was wonderful news to me, but I kept my face sombre in tune with his mood, being not quite sure how to react. I think he thought that I too would be upset to be leaving such hallowed ground after only a short time in post, and that perhaps I would join with him in mounting some resistance. But much as I tried to moderate my excitement with a fitting show of sober reflection, I practically leapt for joy when I left his office. The prospect of returning home to Wiltshire and of getting back into the air after only nine months in the MoD was better than I could ever have hoped for. By then, I had been off flying duties for 4½ years and had begun to wonder whether I'd tried too hard and become too good as a staff officer. I'd begun to fear that I would never be released from it. Moreover, of those 4½ desk-bound years, more than half had entailed my living as a bachelor, sometimes for weeks at a time and more recently, in my current post, having to endure daily commuting back and forth to London. Counting the time commuting, my working-day in DORS had been thirteen hours long, and I knew that I could not tolerate them for long. More importantly, I had been badly absent from my family for far too much time, particularly if I counted all the weeks I had spent away on MRCA work and on those other Buccaneer and Jaguar trials detachments in my first two tours at Boscombe. For all our sakes, I was desperate to live a more normal family life, and I could barely wait for it.

Refresher Training

Before I could take up my new post at Boscombe, this time as CO of the fighter test squadron, I had some serious refresher flying to do. I had not been airborne since the beginning of 1976, and skills that had been second-nature to me then needed to be relearned. Altogether, this training was to take five months of courses in various parts of the country – another five months of weekly commuting long distances by car. But even this would beat sitting behind a desk – and the prospect of returning to a normal family life and doing what I liked best was at last in sight!

My refresher flying began on the Jet Provost at RAF Leeming, in Yorkshire, where I had started my initial flying training all those years before in 1962. The sense of déjà vu was almost overwhelming as I moved in next door to the room in the officers' mess that I had occupied back then and flew the same aircraft type, albeit the slightly higher performance JP Mk 5 this time round rather than the JP Mk4. Even so, after 4½ years off flying duties, I had a lot of catching up – not so much with the basic mechanics of flying the aircraft, which, almost like riding a bicycle, becomes an instinctive skill that you never forget – but more with the air-traffic procedures and with making decisions far enough ahead to take action in good time. Even on the JP, which flew at a relatively pedestrian pace compared to the fast jets I had flown before, my brain struggled to keep up. My rustiness mainly manifested itself in the radio calls that were required, for example, to penetrate controlled airspace or when returning to base to re-join the landing pattern. At first, everything seemed to be happening too fast; I knew I needed to transmit something, perhaps to report my position or to make a request of some kind, but when I pressed the transmit button my mind went blank. I found my tongue getting in the way of my words, my utterances turning into a garbled version of what I really wanted to say. It must have taken a good ten sorties for me to feel back on top of it again, and then only after a lot of amused looks from my instructors.

In February, I moved to RAF Brawdy in Pembrokeshire to fly Hawker Hunters with 79 Squadron, an RAF reserve squadron whose primary peacetime role was training. This was not just a refresher on the type, but also, in effect, an abbreviated repeat of the operational conversion course I had first completed at Chivenor back in 1964. This included a tactical flying refresher, weaponeering, and lots of low-level navigation sorties in formation with my fellow students. Under Rod Dean, the CO of the unit – an old chum from Aden days – I joined a small course of six pilots, all of us destined to progress to Harrier training: three RAF, two USAF, and one RN, all former fast-jet pilots re-treading like me. On finishing the course, we were packed off to RAF Shawbury in Shropshire for two-weeks of helicopter training on the Gazelle to get us used to the particular demands of hovering and vertical take-offs and landings. At first, if we could stay within the airfield boundary in the hover, we were doing well!

I arrived at RAF Wittering, the Harrier OCU, in mid-April. This conversion course followed the familiar format: two weeks of ground school and simulator training before we were allowed anywhere near the real thing. My old friend, Bill Green, my best and favourite student at Oxford UAS and now a flight commander at the Harrier OCU, had assigned himself to me as my instructor and mentor. I had taught him to fly the Chipmunk at Bicester and now he would be teaching me to fly this somewhat more sophisticated machine. He had become an accomplished Harrier pilot by then – one of the 'best of the best' in the RAF – as VSTOL pilots always thought of themselves (not without some justification).

The Harrier

The Harrier GR3 engine was the Pegasus 11 Mk103 with 21,500lbs static thrust, and the aircraft had an empty weight of around 12,500lbs. Even with full internal fuel and two full 100-gallon drop tanks, therefore, its thrust to weight ratio was just about enough to lift it vertically from a standing start. From my first flight, I fell in love with it. I was a promiscuous aviator with many loves, but this was my all-time favourite. It was such a versatile aircraft and so exhilarating to fly. You could do a short take-off with a full load from an improvised strip in a clearing, hidden among the trees, accelerate to 420 knots for a close-air-support mission a hundred miles away, and then return for a vertical or rolling-vertical landing in that same clearing for rearmament and refuelling forty minutes later. This aircraft was much more versatile than the Hunter, which it superseded, which needed a mile or more of defended tarmac from which to operate.

The Harrier would have been an ideal close-air-support aircraft in Aden. Instead of operating from Khormaksar airfield, sixty miles away from the Radfan, it could have been deployed forward with the Army, making communications more direct and responses much quicker – exactly what the aircraft was designed for. But in NATO's northern European theatre, it would have had a hard time against the Warsaw Pact's dense and sophisticated defensive array if the balloon really had gone up, especially in the poor weather of a northern European winter. The Harrier GR3's avionics and displays were advanced in their day, but by the late-1970s they were already steam-age by comparison with Tornado. Without updated navigation avionics and electronic warfare (EW) counter-measures, the Harrier GR3 pilot's high operational workload would have limited his effectiveness severely.

Although the Harrier was generally a delight to fly, the aircraft also had a few foibles in semi-jet borne flight that could catch out a pilot if he didn't pay attention. Control in roll, pitch, and yaw in the hover and in the transition to and from wing-borne flight came from reaction control valves (RCVs) located at the wing-tips, nose, and tail – through which high-pressure air ducted from the engine was directed by the pilot's control stick and rudder pedals. But these RCVs had limited power, so the pilot had to be careful not to let inertial or aerodynamic forces build up beyond his ability counteract them. To make matters worse, the aircraft was directionally unstable at these low airspeeds, due to the lack of airflow around the fin and the destabilising effect of side-forces around the engine air intakes (called intake momentum drag). The combined effect of this low-speed directional instability and limited RCV control power meant that the pilot had to be particularly careful during a deceleration to the hover when approaching the hover across the wind.

At very low semi-jet-borne airspeeds, crosswind effectively becomes side-slip – in other words, the relative airflow does not come from straight ahead as it would in balanced flight, but from one side or the other. Being directionally unstable in this low-airspeed regime, instead of realigning itself into the wind like a weather-cock, the Harrier tends to diverge – that is, to yaw further away from the wind. If the pilot spots this divergence soon enough, he may be able correct it with his rudder pedals (yaw-RCVs) and bring the nose into wind. But this divergence can develop so quickly in a dynamic transition, that an inexperienced pilot might just not realise what is happening in time. This is despite the rudder-pedal shakers and the swinging wind-vane on the nose that are there to warn him.

It is not the side-slip as such that now threatens to kill him, however, it is its secondary effect. As the aircraft slips sideways, the leading wing produces more lift than the trailing wing, creating a rolling moment (roll with yaw) that is particularly strong in the Harrier due to its swept-wing planform. The pilot instinctively (and perhaps even unconsciously) counteracts this rolling moment with opposite lateral stick deflection. But as the aircraft continues to decelerate towards the hover, the side-slip angle increases for a given cross-wind (think of the vector triangle), and thus the rolling moment will also increase. As it does so, more and more lateral stick will need to be applied by the pilot to counteract it. This should be a further warning to the pilot that he is heading into trouble, but he may miss this clue like he has missed the others (he is really not paying attention, is he?). And since at these low airspeeds the roll-RCVs have only limited power, it is quite possible even in moderate cross-winds to run out of lateral control completely. Once this happens, all is lost; the pilot has perhaps a second or two at best to eject before the aircraft rolls to such an extreme angle that his ejection trajectory is downwards (which could give him a bit of a headache).

The key to a successful transition, therefore, was simple – watch the wind-vane on the aircraft's nose like a hawk and ensure that it remains aligned for-and-aft. For all its sophistication in other respects, it was a mechanical device as simple as a wind-vane that kept these early Harrier pilots safe.

The joint US/UK Harrier II (AV8B/GR5/7) was a considerable improvement on the GR3 in all respects. Its greater thrust/weight ratio, bigger wing, lift-improvement devices, automatic stability augmentation, and modern avionics suite brought the aircraft up to date. But the RAF regrettably retired its own in 2011, selling seventy-seven aircraft of its remaining inventory to the US Marine Corps for spares! However, as I write this (in 2020), the RAF and RN eagerly await the arrival of the F-35 Lightning, a third-generation V/STOL aircraft – and so the concept continues, if largely US led. I'm sure that the F-35, with all its technological and performance-enhancing refinements, will be much more capable, operationally effective, and probably also a lot more forgiving in cross-wind transitions than the GR3; but for me and for my generation, the original Harrier was special. It was the first of its kind, and to have had the chance to fly it was a privilege. It endowed us pilots with the agility and capabilities of the bird of prey that gave the aircraft its name. Bill Green, sadly killed in a Tornado accident a decade later, was a great Harrier pilot, a good friend, and an excellent tutor. After eight furious weeks of training with him, he generously pronounced me a competent 'bona-mate', despite my having nearly drowned him in Rutland Water during a squally evening's sailing in my Enterprise.

With this new and coveted qualification signed-off in my log-book, I was sent back to Manching (the MBB flight-test centre in Bavaria) for the final course of my refresher programme, this time a three-week technical update on Tornado. The production version of the new aircraft was by then being tested at Boscombe, and I would be re-joining the flight-test team. Once I had been duly updated, I would at last be ready to start my new job as OC 'A' Squadron, which was perhaps the most coveted of all military test flying jobs in the UK at that time (but then I would say that, wouldn't I?). Those past five months had been a busy and intensive grand tour, but I was now well and truly back up to speed, and oh what a relief it was to be back in the air!

CHAPTER 15

Officer Commanding 'A' (Fighter Test) Squadron

I took over the command of the fighter test squadron from David Scouller on 15 June 1981 and found myself once more working for my old boss and NAMMA colleague, Group Captain Ned Frith, Superintendent of the Test Flying and Flight Test Training Division. The Squadron was still in more or less the same shape and size that I had left it some five years previously, but with new faces and new projects. It still occupied the hangar and office buildings at the far north-eastern end of the line where it had been for decades, but we were soon to be relocated to the larger and more centrally placed former 'B' Squadron hangar, which had more room for our expanded inventory. 'B' Squadron had by then moved into the larger Weighbridge hangar, having absorbed the aircraft and testing roles of the disbanded 'E' (Transport Test) Squadron. Our aircraft included Harrier GR3 and Sea Harrier FRS1 (the naval version of the Harrier), Tornado GR1, Jaguar, Buccaneer, and Phantom, but we also flew Hawk, Hunter, and Jet Provost from the training fleet, and were soon to receive the Tornado air defence variant (ADV) for its early clearance trials. While I regularly flew all the types listed above, sometimes testing new or modified equipment, sometimes as photo-chase for other trials, and sometimes for continuity training, most of my trials flying would be on Tornado and Harrier.

Tornado testing had moved on a lot since my last hands-on experience in prototype P-01 in 1975, and clearance trials on production models were now well underway. I got my checkout on production Tornado XV631 with Graham Tomlinson on 17 September. It was like I'd never been away from it. The prototype's teething problems – the 'vertical bounce' and the PIO tendencies referred to earlier – had been sorted out, and instead of P-01's bare prototype cockpit, 631 was fitted with the full avionic suite and cockpit layout on which I had spent so many years working in the early 1970s. I felt instantly at home in the aircraft and it was a pleasure to fly such a potent and capable machine, especially one for which I had developed a proprietorial affection. The Head-up display, moreover, now incorporated all the symbology improvements that my NAMMA HUD improvement programme with Don Thomas had successfully developed.

I re-entered the programme as it was beginning certification trials of the aircraft's various external stores configurations. The Tornado carried all its stores externally (unlike the Buccaneer, which could carry four one-thousand pounders inside its rotating bomb bay), and each different weapon configuration could potentially present different handling and performance

125

characteristics. For example, one of the war-load configurations included eight 1,000lb bombs on the fuselage stations, two 1,500 litre external fuel tanks on the inner wing pylons with air-to-air missiles on the 'shoulder' stations, and two electronic counter-measures pods on the outboard wing pylons. In other configurations, Tornado might carry variations of the above in different positions, or carry different armament with different shapes and weights, or up to four cruise missiles, or the large JP233 airfield attack weapon on the centreline station. Moreover, as weapons or stores were released, say during a bombing attack or if jettisoned, a partial or asymmetric configuration might be left behind – different again from the handling and performance point of view. A sufficient number of these configurations had therefore to be tested to establish safe limits of control across the whole flight envelope, as well as to quantify the performance impact of each on such parameters as drag, fuel consumption, and thus mission radius. Add to all this, the different wing sweep configurations used by the aircraft operationally between 25° and 68°, and you might get some idea of the size of the flight test clearance programme needed to be confident that all the possible permutations had been adequately covered. We were not just testing one aircraft, we were effectively testing several different aircraft because of the wide spectrum of aerodynamic shapes, weights, and centres of gravity. This is where the CSAS really came into its own, compensating for these variations by adjusting flight control responses independently of the pilot's inputs, so that the aircraft's apparent flying qualities remained reasonably consistent throughout the envelope. This could only have been achieved with a computerised fly-by-wire flight-control system. Conventional flight-control systems would not have been able to compensate effectively for such a wide spectrum of configurations, some of which would otherwise have presented very different handling qualities and limits of safe control.

On top of this, the release and/or jettison of these stores across the range of possible wing-sweeps, airspeeds, and Mach numbers also had to be tested to establish safe release criteria as well as to quantify the effect of post-release airflow disturbances on weapon trajectories and thus weapon-aiming accuracy. There was thus a huge amount of work to do in this respect, which meant scores of weapon and stores releases over the Larkhill, West Freugh, and Cardigan Bay ranges up to supersonic speeds, using ground-based photographic tracking facilities as well as our own on-board photography and photographic chase aircraft.

The Tornado's prime operational mission was all-weather, low-level, interdiction-strike in an environment where a great array of anti-aircraft guided or radar-aimed weapons may be deployed against it. The testing and calibration of those of the Tornado's systems that had been designed to make the aircraft effective in such environments was thus also part of our job. This meant placing the aircraft in situations which as closely as possible replicated the operational environment so as to evaluate how effectively those systems performed. And so I soon found myself flying the Tornado at low level in the highlands of Scotland at night and in patchy cloud, monitoring the automatic terrain-following system as it did its job, while the navigator in the rear-seat steered us with his radar towards our simulated target. When I say that I was flying the aircraft, I really mean that I was sitting in it with my eyes glued to the terrain-following radar screen and my hands hovering near the stick so that I could take over manually if the system misbehaved. I remember being troubled by the lights that occasionally flashed past above us on these fast, blind runs down Scottish valleys, only to be reassured by my navigator that they were the lighted windows of isolated houses half-way up the hill-sides! Interestingly, we got used to this quite quickly and learned to trust the system and relax (if not too much). 'Why die all tensed up?' we used to say (with more than a little false bravado).

This is perhaps a good point to pay tribute to our rear-seaters, specialist navigators like Bob Ross, Peter Huett, and Barry Hardy-Gillings who were not only doing testing of their own on rear-cockpit equipment, but were also enduring these wild rides of ours without the comfort of a front-seat view. Imagine sitting in the back seat of a performance car being driven at night at high speed down rough and winding country lanes while trying to read a map by torchlight, and you might get some idea of what it must have felt like.

The automatic terrain-following system had all sorts of safety features in its circuitry, which would roll the wings level and pull the aircraft up into a climb should anything go wrong – but, of course, these safety features needed to be tested too. That the aircraft had these safeguards was reassuring, but an emergency pull-up in hostile territory risked putting the aircraft within sight of enemy radar and so it could be a real case of 'out of the frying pan and into the fire'. Thus, in our testing, we set out to expose the aircraft to the worst situations likely to be encountered operationally in order to satisfy ourselves that the system would not trigger a pull-up too often.

Perhaps the most demanding of these situations involved flying directly towards tall masts and power cables, where the narrow cross-section of the structure or cables might not be detected by the radar until quite late on in the approach. We would therefore set up test routes that included such features, and observe how the system coped, always ready to intervene if necessary. Apart from the obvious collision risk, an important operational concern was that late detection of such a hazard, and a consequent last-second and over-steep command for the aircraft to climb out of harm's way, would cause it to pitch up much higher than necessary and thus make itself vulnerable. The ideal flight path would be to circumnavigate the feature completely if seen in time, or if that were not possible, to ascend early and thus not so steeply. In those Cold War days, we assumed that barrage balloons with hanging cables would be deployed defensively around prime targets, so this testing was not merely for academic interest.

Operation Corporate – The Falklands Conflict

As far as Harrier trials flying was concerned, my own involvement during my first year as CO was limited to taking part in some relatively mundane navigation and radio trials. Practically all of the original clearance work on the aircraft had long since been completed, and so any trials flying that 'A' Squadron was asked to carry out in 1981 was related to new equipment or the investigation of problems that had arisen during in-service use. Then in April 1982, Argentina decided to invade the Falkland Islands, South Georgia and the South Sandwich Islands – three British dependent territories in the South Atlantic. The British Government's response was to send a task force that was eventually to comprise 127 vessels: forty-three Royal Navy vessels including two aircraft carriers: HMS *Invincible* and HMS *Hermes*; twenty-two Royal Fleet Auxiliary ships; and sixty-two merchant ships, which between them (inter alia) carried amphibious and land forces and associated logistic and other support units including RN and RAF helicopters. On board the carriers were twenty-eight RN Sea Harriers and fourteen RAF Harrier GR3s, which would be pitched against the 122 fighters and strike aircraft known to be available to the Argentinian forces. Other RAF resources utilised in the campaign would include almost their entire Victor K2 flight-refuelling tanker fleet, which was necessary to support the single 8,000-mile round-trip Vulcan bombing raid on Stanley airfield on 1 May. I mention all this to set the scene, because it is relevant to the work in which 'A' Squadron and I subsequently became involved.

As the task force steamed south, urgent requests were received from both the RN and the RAF operations staffs in the MOD for additional flight test work to be carried out on the Harrier. This was to include clearance recommendations for the carriage and release of additional stores and weapons on both versions of Harrier (including Sidewinder on the Harrier GR3), Harrier GR3 take-offs using the 12° and 7° 'ski-jump' ramps fitted to HMS *Hermes* and HMS *Invincible* respectively, and air-to-air refuelling of the Harrier from the newly modified C-130 Hercules Mk1K air-to-air refuelling tanker. 'A' Squadron had lost one of its Harrier test pilots, Lieutenant Commander Rod Fredrickson, who had been called up within days of the Argentinian invasion to re-join his squadron aboard HMS *Invincible*. This left me with two: my old friend and fellow USNTPS graduate, Lieutenant Commander David Poole RN, who had remained behind to carry out additional Sea Harrier trials on the ski-jump at RNAS Yeovilton, and Squadron Leader Tim Allen, a former RAF operational Harrier pilot. With these two experienced VSTOL test pilots doing most of the Harrier work, no contribution from me was initially required. But events were soon to intervene.

David was carrying out a series of take-offs in Sea Harrier XZ438 in a variety of configurations from the ski-jump at RNAS Yeovilton when an asymmetric fuel load condition developed which caused his aircraft to roll uncontrollably at the point of launch. I've already mentioned the limited roll-control power of the early Harrier RCVs that could lead to trouble in semi-jet-borne flight. David's situation represented an extreme case of this – not this time caused by crosswind, but by asymmetric weight. He just didn't have enough roll control power at launch airspeed to correct the out-of-balance load. Fortunately, he recognised this quickly and managed to eject just before the aircraft rolled over on its back, but his ejection trajectory was very shallow and the parachute had only the barest amount of time to open before he hit the ground. Rod Job, the A&AEE trials officer running the ski-jump trials at Yeovilton, watched the whole thing from alongside the ski-jump ramp. He recalls from his notes that the aircraft was already descending at 1,200ft per minute with a bank angle increasing through 60° at the point of David's ejection. The Martin-Baker Mk10 seat on which David sat was rated to be able to eject the pilot safely at zero height, but the aircraft's descent rate would have added about 120ft to this (10 per cent of the rate of descent). At only 80ft above the ground at the point of his ejection, therefore, David was very lucky that the seat performed better than advertised.

Although not seriously hurt, David sustained some minor injuries that were to put him out of action for a while, which meant that 'A' Squadron was now two Harrier test pilots short. And so, because of the urgency and the amount of work remaining to be done, I, as the only other qualified Harrier test pilot on the Squadron, had rapidly to gird my loins!

Chapter 16

Ski Jumping the Harrier

After an accelerated Harrier refresher with Bill Green at Wittering, and a couple of practice take-offs on the ski-jump in the two-seater Harrier T4 with Tim Allen, I was ready to go. Although the notion of being thrown into the air without enough airspeed to sustain flight was a little scary at first, ski-jump take-offs proved to be the easiest way of getting the Harrier airborne. Using this method of launch, the aircraft was literally thrown up into the air by the ramp, giving it greater vertical separation from the ground and thus the pilot more time and height to jettison stores or eject if things went wrong. The aircraft also became airborne without the usual ground-effect trim changes that sometimes took unwary pilots by surprise when taking off from a runway. Launched directly into the air at the top of the ramp, the aircraft had already been pitched into a healthy nose-up attitude at the right angle-of-attack, so there was not much actual controlling to do. In fact, with the right trim setting, you could probably do it hands-off (although I never did). We spoke of 'gathering' or 'collecting' the aircraft as it attained flying speed at the apogee of its partly jet-borne, partly wing-borne, and partly ballistic trajectory – because that is exactly how it felt. It was like gathering the reins of your mount having given it its head for the jump.

Perhaps the most critical factor in the use of the ski-jump was the calculation of the position on the deck from which you would commence your take-off run. The length of the run would depend on the aircraft's weight and configuration, the wind-speed over the deck, centre-of-gravity, and the engine thrust (taking air density and air temperature into account). If you did your calculations properly (using a carpet graph) and started your run from the correct position, you would enter the ski-jump slope at the right speed to ensure a healthy leap into the air with an easy transition to wing-borne flight. Too fast, and the undercarriage might be damaged due to excessive compression forces; too slow, and you risked an undignified descent into the sea and being run over by your mother ship. Just as your wheels came to the top of the ramp, with your nose now pointing heavenwards, the nozzle lever would be pulled back to a pre-set stop, usually set at about 30° nozzle-down depending upon conditions. While still providing enough forward thrust to continue the aircraft's longitudinal acceleration, this downward deflection of jet efflux would also create an upward force that would supplement wing lift. If the aircraft had arrived at the lip of the ski-jump at the right airspeed, the upward momentum achieved, aided by this jet-lift component, would ensure a good trajectory and enough time and space to accelerate to wing-borne flying airspeed before gravity took over. The pilot would then 'gather' the controls, ease the nozzle lever gently into the fully-forward position (nozzles fully aft), and fly away in conventional wing-borne flight, raising undercarriage and flaps as airspeed increased.

I carried out a lot of ski jumping at Yeovilton in the GR3 at the two ramp angles (7° and 12°), and at different weights and in different configurations under Rod Job's watchful eye. Some of these test points were quite new because the RAF had asked us to clear the carriage of several new stores, including Sidewinder air-to-air missiles and special containers for equipment delivery. Tim Allen and I split the work between us and we completed all the tasks required of us in good time for our clearances and recommendations to prove useful in the South Atlantic.

The next task was to clear the Harrier for air-to-air refuelling from the C-130 Hercules Mk1K air-to-air (AAR) refuelling tanker. Six of these transport aircraft had been newly modified with a centre-line hose and drogue unit to supplement the RAF's AAR fleet, which had become severely stretched in supporting air-defence operations and supply routes in the South Atlantic. I had completed AAR ground school training only a few days before the Falklands invasion took place, only because it was our intention to use 'A' Squadron's Buccaneer XV337 as a mini-tanker to support our Tornado test programme. It was purely fortuitous, therefore, that I was AAR 'qualified' when the Hercules tanker arrived at Boscombe. But having only refuelled a few times from a Victor tanker in the North Sea training area, I could hardly have called myself experienced in the role. Within a few weeks, however, I had become a veteran. In fairly rapid succession, I carried out numerous AAR clearance trials flights against the Hercules and other tankers flying Harrier, Tornado and Buccaneer, all without problems. Later that same year, clearance trials on the new Vulcan B2(K) tankers began. Six of these venerable 'V' bombers had also been modified as tankers by fitting a single hose-drum unit (HODU) in the tail cone. This hurried modification had been another response to the continuing need for AAR in supporting long-range supply flights to the South Atlantic after the end of the Falklands' war. It was a stop-gap measure until the new VC10 tanker came on line.

The Vulcan AAR modification featured a very inelegant box-like housing about the size of a cabin trunk, crudely riveted into place; but despite this ugly protrusion the aircraft produced by far the smoothest refuelling environment I had ever experienced. The wing-tip vortices were well away from the centreline and so did not interfere, and the smooth underbelly of the delta-wing planform produced almost no turbulence. It was such a clean airflow environment that the drogue (into which you had to insert your refuelling probe to receive fuel) remained almost rock-steady as you approached and made contact. Indeed, making contact was so easy in the Harrier that I was able to complete all the required test points in only two sorties, and no adjustments to the HODU were necessary. The first converted B2(K) Vulcan flew on 18 June 1982 and entered service with No 50 Squadron only six days later. No. 50 Squadron was the last unit to operate the Vulcan until it disbanded in March 1984.

At the other end of the scale of handling difficulty was Tornado versus Victor K2, particularly when refuelling from an outboard wing station. The Victor's wing-tips produced strong vortices, and its underwing pods and external fuel tanks created additional airflow disturbances. Combined with the bow-wave effect produced by the approaching Tornado's rather blunt radar nose-cone, the hose and drogue behaviour could become a little unpredictable. Successful contact could be achieved most of the time, but it was generally much harder work than in the Harrier, and sometimes, literally, a hit-and-miss affair, particularly if air turbulence was present. If you can imagine a bar-magnet dangling on a piece of string, north-pole downwards, and bringing another bar-magnet up to it from below, north-pole upwards, then you get some idea what it sometimes felt like. As your probe approached the basket, it moved away as though repelled. You had to learn to anticipate this itinerant behaviour and capture the basket before it made a run for it! If you didn't nail it the first time, you'd have to back off, let the hose settle down, and then come in again for another try.

The following year, when the new VC10 tanker and Tornado F2 air-defence variant (ADV) arrived at Boscombe for clearance trials, I carried out some of the AAR work on those too. I remember one sortie in particular that had been arranged to show off these two new aircraft to a visiting

CHAPTER 16

defence-college group at RAF Brize Norton. We were asked to perform a fly-past with the ADV coupled up to the VC10 tanker. It was a rather grim day as I remember, with scattered cloud and gusty wind. I had just carried out a supersonic test firing of the Skyflash air-to-air missile in the Aberporth range area over Cardigan Bay and had planned to rendezvous with the VC10 on my way back. Normally, flight refuelling would take place at a medium or high altitude where there was the least air turbulence, but I was vectored onto the VC10 at 5,000ft as it descended into Brize for the fly-past. I caught up the tanker just below cloud in clear but turbulent air. The drogue was bouncing around all over the place as I manoeuvred into position behind it, and much as I sweated over the task to get my probe into phase as I followed the VC10 down onto the display line, there was no way I could safely make contact in those conditions. Our audience undoubtedly noticed that my probe was a couple of feet aft of the bouncing drogue as we roared along the runway at 1,000ft, but it must have been a pretty impressive sight nevertheless.

A word or two in passing about the Tornado air-defence variant, which was for me (and many others at the time, I think) a bit of disappointment, both from the handling and the performance points of view, although no great surprise I have to say. The aircraft was never an air-superiority fighter, nor could it have been; it simply lacked the thrust and wing to make it agile enough in manoeuvre to engage head-to-head with the real fighters of the day. 'Multi-role' in the aircraft's original definition (MRCA) was more related to the different low-level roles such as interdiction, strike, maritime, and reconnaissance, and so it was a bit of a leap of the imagination to give it the prefix 'F' for fighter. 'I' for interceptor might have been more accurate. Germany and Italy must have recognised this for they never ordered it. The Tornado F2 flew exactly like you might expect given its original interdiction/strike heritage: a bomber or heavy attack aircraft. Although it was made lighter on the controls than the IDS version, it was a pig to manoeuvre at high altitude and spent a lot of its time in buffet.

Compared to the IDS, the ADV wing gloves were swept at a greater angle to reduce wave drag, and the Kruger flaps and the port Mauser cannon were removed to reduce weight. The aircraft also had a longer radome to accommodate the Foxhunter air-to-air radar, slightly longer air-brakes, and a 1½ metre stretch of the fuselage to allow for the carriage of four Skyflash semi-active radar-homing missiles. The stretch was applied to the front-fuselage immediately behind the cockpit and had the unexpected benefit of reducing drag as well as making space for an additional fuel tank. The extra fuselage length also made it look a little less dumpy than the rather squat IDS version, and in F2 prototype ZA254's sleek black-and-white livery, it really looked the business when it was delivered to 'A' Squadron in April 1982, even though it never fully delivered on its promise.

The Tornado F3 was certainly an improvement, although unfortunately I never got to fly it. Its engines were better optimised for high-altitude use and its airframe modified to carry four underwing Sidewinder missiles rather than the two carried on the F2. Automatic wing sweep control (rather than manual) was also incorporated as an aid to manoeuvring, so this should have helped a bit in one-to-one combat situations if ever encountered. Perhaps it is best not to dwell too much on the ADV's early air-intercept radar performance. Some readers may remember the radar's development problems which, early in the aircraft's life, necessitated the carriage of concrete ballast in the nose to compensate for its absence – labelled the 'Blue Circle' radar by those who knew. Nevertheless, the aircraft had the ability to stay aloft for long periods, and the performance of its weapon systems was also a dramatic improvement over its predecessors, the Lightning F6 and F4 Phantom. Compared with the Phantom, the Tornado ADV had greater acceleration, twice the range and loiter time, and was more capable of operating from short or damaged air strips. Previous interceptors, moreover, were reliant on ground-based radar stations, while the F3's radar was (eventually) capable of performing much longer-range and wider scans of surrounding airspace, tracking and engaging targets at far greater distances. The Tornado also had the ability to share its radar and targeting information with other aircraft via JTIDS (Joint Tactical Information Distribution System). With various upgrades, the ADV would remain in RAF service for twenty years until 2011.

In-Flight Refuelling

Returning now to the theme of in-flight refuelling... A tanker's refuelling hoses are stored on hydraulically powered drums within its fuselage or wing pods and unwound when in use so that the drogue (or basket as it is often called) sits between about 60 to 90ft behind the tanker. The hose is kept full of fuel so as to add weight and stiffness, which help to dampen any skittishness or instabilities arising from air turbulence. In this extended position, the air drag on the drogue is balanced by the take-up torque of the drum's hydraulic motor so that the drogue remains steady in the correct fore-and-aft position. It is the setting-up of this take-up torque that becomes quite critical in initial flight testing because, until actual flight experience is gained, the adjustment of torque is based only on mathematical modelling. Too little torque, and making contact with the drogue will feel like receiving a punch in the nose; while too much torque will cause the drogue to retreat from the probe like a rabbit into its hole the moment it feels the slightest caress of the receiving aircraft's probe.

The drogue itself is a little less than 2ft in diameter and shaped like a shuttlecock with metal spokes that help to guide the receiver's probe into the fuel valve located at its centre. This is not as easy as it sounds; the drogue can be frisky and elusive, moved around not only by air turbulence but also by the approaching receiver's bow wave, as has already been mentioned. Moreover, there is at first a psychological aversion to making physical contact with another aircraft which, in other circumstances, would usually result in a board of enquiry! Introductory in-flight refuelling sorties are thus quite tense affairs, often followed by a shower and a change of underwear. In a normal in-flight refuelling session, the tanker is held steady at an airspeed of between 200 knots and 350 knots (depending upon the tanker and receiving aircraft), and the receiver approaches the drogue with an overtake speed of about 3 knots – equivalent to a good walking pace – so as to create a positive connection between the probe and the fuel valve. If the approach is too slow, the fuel-valve's spring-loaded mechanism may not close properly around the probe, making the connection insecure and causing fuel to spill from the valve. If the approach is too fast, especially if the probe enters the basket off-centre, the impact could break the probe off or damage the drogue. The probe could even push its way through the spokes and get stuck, after which something will have to give!

Fortunately, probes and drogues have weak-link attachment points built in, so that if the pilot is being a bit clumsy, it would be these weak-links that would break rather than something more critical. Causing a breakage like this, however, would not be good for your reputation – taxiing back to your squadron dispersal minus a receiving nozzle, or worse, with the tanker's drogue perched on the end of your probe, would draw a few mocking looks from your crew-room colleagues – and you'd also have some explaining to do. But in reasonable refuelling conditions and with a bit of practice, most pilots could expect to make a good, firm connection first time, most of the time.

Making contact is just the beginning, however. Once engaged, the receiving pilot then has to push the drogue forwards by about 10ft (assisted by the hose-drum unit take-up torque), at which point the tanker's fuel valve will open and permit fuel to flow (at about 2,000 litres per minute). Job done! All the receiving pilot has to do now is maintain this fore-and-aft position plus or minus 10ft or so until the required amount of fuel is taken on board. In comparison with the workload required actually to make contact, this is the easy bit, even when the tanker needs to manoeuvre – it's just formation flying in close line-astern, one of the easiest positions to fly.

In all my AAR testing, I didn't encounter many problems, but one experience stands out in my memory. This occurred during the first test of a new hose and drogue system, and was caused by the hose-drum take-up torque initially being set too low by the manufacturer. As previously described, too low a torque setting meant that there was no take-up or 'give' in the hose as contact was made, which made it feel a bit like I had just walked face-first into a lamp post (which I did once, and so I know). The shock of the impact created a travelling wave in the hose much like you would get by flicking a skipping rope. And just like a rope tied to a wall, the wave travelled up the hose to the tanker and then bounced back towards me, seeming to gather pace and amplitude as it did so. The photo-chase pilot said that it all happened in a very few seconds, but for me, it felt much longer as I watched the wave first retreat towards the tanker and then bounce back with a vengeance. I knew immediately what I was in for, but I could not get out of the way quickly enough. 'Uh oh … ' I uttered into the microphone for the navigator's benefit, and braced myself. The returning wave arrived back like an express train and tore off the end of my probe. The hose and drogue, now freed from the restraining attachment to my probe, at once began to flail about wildly with my probe-end stuck firmly in its fuel valve, threatening to box me about the ears and do even more damage. By this time, however, I had already closed my throttles and was retreating rapidly, while also moving off to the side in case the severed probe loosened and came back at me like a cannonball. It didn't, thank goodness, but after this, both the tanker crew and I were forced to retire for an early tea. Adjustments were later made to the take-up torque to resolve the problem, but you can imagine the consequences of such an event had it happened half way across the Atlantic.

Chaff

Harrier operations in the Falklands War also highlighted the need to improve its survival chances when fired at by air-to-air and ground-to-air missiles and radar-laid anti-aircraft guns. Military aircraft going right back to those of the Second World War have used chaff or 'window' to confuse enemy radar, and the Americans used it extensively in the Vietnam war. Surprisingly, modern British aircraft, right up until the late 1970s were not routinely equipped with any means of dispensing it. Even at the time of hostilities in the Falklands in 1982, the Harrier lacked any radar counter-measures. Only as the fleet steamed southwards was the Harrier's vulnerability addressed. RN engineers hurriedly improvised a system of dispensing chaff by storing packets of the radar-reflective material in the air brake. These packets would be released by a Heath-Robinson arrangement of wires and split pins when the air brake was cracked open using the pilot's cockpit switch. This must have offered some reassurance to British pilots overflying missile-equipped Argentinian troops or when tangling with an Argentinian Mirage, but it was a temporary one-shot lash-up of limited effectiveness and unsuitable as a permanent facility.

In the August of that year, therefore, 'A' Squadron was tasked to evaluate the effectiveness of the new BAe Phimat chaff-dispensing pod on the Harrier as a means of improving its survivability against such threats. With both 'A' Squadron's RN Harrier pilots still away and Tim Allen employed on Sidewinder and other trials, I very happily took this work on. Over eight days, I flew seven sorties trying out different chaff delivery profiles to establish those that would give the best levels of protection. This entailed flying at very low altitude against simulated targets on the nearby Larkhill range while an array of ground-to-air radars attempted to lock on to me. Our first important conclusion was that you could not depend upon chaff alone, even using ground contours for cover. Tracking radars might be tricked momentarily to lock on the little puffs of reflecting material ejected in the aircraft's wake, but modern radars were programmed not to be so easily fooled. The chaff cloud could not possibly be an aircraft since, with no momentum of its own, it would stop abruptly in mid-air almost as quickly as it had been ejected, something that even the Harrier could not do. The radar's computer memory used earlier tracking data to predict the forward flight-path of its target, and with the decoy chaff cloud seen immediately for what it was, it was ignored. The radar head now simply motored fast-forward along the computed flight-path until the real target was reacquired. Having realised that this was the case after the first trials sortie, our test programme thereafter focused on developing manoeuvring tactics that would not make the aircraft's flight profile quite so predictable ... and the least said about that, the better, I think.

An Argentinian Perspective of the Falkland Islands' War

The Falkland Islands conflict was code-named Operation Corporate in the UK, and known as La Guerre de las Malvinas in Argentina. In 1992, on the tenth anniversary of the start of this undeclared war, I would find myself hosted by the Fuerza Aérea Argentina (FAA) at the Argentinian Flight Test Centre in Cordoba, invited there by Comodoro (Colonel) Juan Carlos Sapolski, the then Argentinian chief test pilot. I was by then the Principal and Director of Training of the International Test Pilots' School, and was on a sales tour of South America, hoping to enrol some of his test aircrew onto our courses. It was an odd experience to be among some of the Argentinian aircrew who had been on the opposite side of the war only ten years previously and yet to be made to feel so welcome as a fellow military aviator. Nationalities put aside, we were all brothers-in-arms, so to speak, professional fighting men who obeyed the call to duty when it came. Having flown the Harrier, the A4 Skyhawk, and the Argentinian twin turbo-prop ground-attack Pucará, I was in the unusual position of being able to compare these aircraft in their respective roles. In their crew-room, I fell into conversation with a dozen or so Argentinian pilots debating the different tactics employed. Their perspective on the war was illuminating and I was privileged to gain some insights that may be worth sharing.

They would say that the A4 Skyhawk would have been a good match for the Harrier in manoeuvring flight if respective firepower and weapon loads had been equivalent, and I would agree. But, heavily-laden with bombs as they ran in for their attacks, their A4s were slower and less agile than the Harrier. And operating at such long range from their bases on the mainland, they did not have the fuel to hang around for a dog-fight if they were set upon. Their attack profiles were thus essentially straight-in and straight-out and, although flying very low as fast they could, they knew that they were vulnerable once they had been spotted. The Sea Harriers, alerted of their approach to the Islands by airborne and ship-borne radars, were able to get in behind and launch their Sidewinder missiles with deadly effect. Several Argentinian Mirages and Dagger fighters sent in as top-cover for the Skyhawks were also shot down in this way. Had range and fuel not been an issue for the Argentinian aircraft – if they could have hung around to draw the Harrier into air-to-air combat – things might have turned out differently. Certainly, more Harriers would have been lost. But while the Harrier's turning performance may not have been as good as its potential adversaries, it had two advantages that would anyway have levelled the odds.

First, the Harrier's high thrust-to-weight ratio would have allowed it to preserve manoeuvring energy quite well in a protracted turning fight, and also give it rapid acceleration if it needed to break out of a one-to-one tussle and make a run for it. Secondly, its vectored thrust could be used tactically both offensively and defensively. In a dog-fight the winner will be the one who gets behind the other and obtains a firing solution for his guns or missiles. For a guns attack, the pilot also needs to point his guns sufficiently ahead of his target to allow for gravity-drop and the flight-time of the projectiles. This is called 'lead' (familiar to pheasant shooters); but with both target and attacking aircraft manoeuvring at the limits of their turning capability, it may not always be possible to get enough of it (because of wing-stalling). Even using a Sidewinder air-to-air heat-seeking missile, the attacker still has to bring the target into the missile's relatively limited 'field of view'. This is where thrust vectoring (VIFFing– vectoring in forward flight) could help the Harrier, by supplying a bit of extra lift (vertical component of thrust), which momentarily allowed the pilot to inch his nose ahead of his target for just long enough to obtain a firing solution.

If, on the other hand, the Harrier pilot finds himself in someone else's sights (heaven forbid!), he could throw his nozzles into reverse (in the extreme) and force his attacker to fly through (overtake). If, then, he was quick about it, the Harrier pilot may then be able to loose off a passing shot or two as his surprised assailant thunders past (wondering what the hell just happened!). But the disadvantage of both these VIFF tactics is that energy is given up – in the second case, practically all of it. This now puts the Harrier in a very vulnerable position if his assailant has a number two coming up behind to make the Harrier the meat in the sandwich. VIFFing sounds a great idea, but by giving up his speed in using the technique, the Harrier pilot has to be careful not to make himself a sitting duck.

Some of the early Argentinian losses were put down to lack of combat experience, but with a four-to-one advantage in combat aircraft numbers, the Argentinian MoD believed they would eventually turn the tables through attrition. Sea Harrier numbers would soon thin out through losses, they thought, and this inevitably would force a less well-defended British fleet to pull back. But then came the RAF Vulcan raid on Port Stanley airfield, which put a hole in its runway. The raid caught the Argentine Government by surprise. The runway at Stanley was only 4,100ft long and unsuitable for jet attack and fighter aircraft, but the bomb crater prevented its use even by the Puccara and by supply aircraft from the mainland (for a time). Until repaired, resupply could only take place by sea, which was seen as extremely hazardous considering the UK's maritime forces that stood in the way. But perhaps more importantly, the Argentinians realised that if Stanley airfield could be bombed, then so could Buenos Aires.

It was this fear of attack on the Argentinian capital that led to the redeployment of their air-defence resources, which thereafter were held back to protect the mainland. The repercussions of this change of focus meant that airborne attacks on the British fleet now had less fighter 'top-cover' making the A4s yet more vulnerable, but it also also severely compromised their earlier strategy of winning air superiority through attrition. Then came the sinking of the Argentinian battleship, *General Belgrano*, by a British submarine, with tragic loss of Argentinian life. Almost overnight, the entire Argentinian fleet, including the aircraft carrier (*Vienticinco de Mayo*) that had been the launch platform for the S2 Trackers and some of the Skyhawks, was called back to port for protection. Those vessels never returned to the fight. The tables had thus been turned, and from that point on, the eventual outcome became predictable.

That wretched war has been analysed often enough and it is inappropriate for me to offer further comment here, but I found myself admiring the FAA pilots for their tenacity and courage in the face of superior British equipment, weaponry, and combat skills. I spent several days with Juan Carlos and the pilots of his test squadron, including a barbecue with his family and a memorable squadron dinner in a Cordoba steak house. He also gave up a day to show me the surrounding countryside in the foothills of the Andes. I left Argentina feeling that I had made a good friend. After returning to the UK, I was very pleased to hear that his squadron had signed up for some training courses with us back at Cranfield later in the year and I looked forward to returning the hospitality that I had received in Cordoba. Juan Carlos himself and Captain Omar Gelardi, one of his pilots whom I had also got to know well on my visit, brought their new Argentinian Pampa IA-63 advanced trainer to the UK for demonstration at the 1992 Farnborough Air-show. They based it at Bournemouth airport to work up and rehearse their display. Living not far away near Salisbury, I had planned to meet them for dinner one evening, but sadly this was not to be. On the morning of 31 August, while practising their aerobatic routine over Hurn airfield, the aircraft failed to pull out of a looping manoeuvre and hit the ground at high speed, killing both men instantly. It appears that they had both initiated the ejection sequence, but it was too late to save them. Juan Carlos and Omar were the last of a dozen in my own circle of good friends and colleagues to be killed in aircraft accidents during my thirty-years in military aviation.

CHAPTER 16

Stores' Release

With the successful retaking of the Falkland Islands and the completion of trials specifically in support of that operation, 'A' Squadron's work on routine clearance trials resumed. Foremost of these as far as I was concerned were the continuation of Tornado terrain-following testing and trials of high-speed weapon aiming and stores release. Enough has already been said about the Tornado's terrain following system, but the release of weapons, and indeed the release or jettison of any stores from an aircraft, especially at high speed, was more potentially hazardous than it may at first seem. Anything released by an aircraft in flight becomes a flying object in its own right, and like stocks and shares, such articles can go up as well as down once released from their constraint. From modern military aircraft, stores such as weapons, fuel tanks, counter-measures pods, and missiles, are not simply released from their carriage mechanism and allowed to drop under the influence of gravity, they are forcibly ejected by front and rear pistons powered by springs or explosive cartridges.

The key to successful release is to ensure that in departing from the aircraft, these objects are set immediately to an appropriate angle of attack to produce a flight path that is both safe for the releasing aircraft as well as predictable and consistent. This is achieved by adjusting the forces of these pistons or springs accordingly. If the attitude of the store on ejection is too nose-up, stores will develop upward lift that could result in collisions with the aircraft, possibly with catastrophic effect. If the attitude on release is too nose-down, the stores could pitch down or tumble, striking other stores being released at the same time. For the jettison of inert stores such as ECM pods or fuel tanks, glancing collisions on the way down may not matter too much once the stores are at a safe distance from the launching aircraft. But collisions between explosive weapons immediately underneath the launching aircraft would be another matter entirely. In either case, the store's trajectory after release becomes unpredictable and thus so does the profile of it's subsequent flightpath to earth, making it impossible to predict with any certainty where it will end up. This is not very satisfactory if your objective is pin-point accuracy.

With up to twelve weapon carriage points on the fuselage and inboard-wing pylons of the Tornado, and two more outboard-wing carriage points for counter-measures pods, there were a lot of combinations to clear for service use. This work, like most operationally-related testing and evaluation of new aircraft and weapon systems, was generally allocated to the engineers and military test aircrew of Boscombe Down, rather than to those of the aircraft manufacturer. In this way, Boscombe also ensured that the military services were getting what they had specified, and were not simply (perhaps naively) accepting the manufacturer's assertions. Our programme for the Mach 2+ Tornado, had to include stores release and jettison over a large envelope of speeds and wing-sweep configurations, ranging from relatively low-speed jettisons to supersonic weapon releases, and some Mach 2 missile firings on the ADV, which all needed to be instrumented, photographed and analysed. This was all very interesting and quite challenging work, not only for the Tornado's test aircrew and chase-aircraft pilots, but also for the engineers of the Armament and Navigation and Systems Divisions at Boscombe, with whom we worked very closely.

When I was not carrying out trials flying on Harrier and Tornado, I contributed to the continuing trials on Buccaneer and its new external load configurations, as well as flying photographic chase sorties on Hunter and Hawk. But when there were gaps in these trials programmes, as there often were, I would look for opportunities to fly whatever else I could lay my hands on, so as to stay in reasonable practice. Boscombe kept a small fleet of training aircraft for just this sort of continuity flying – Hunter, Hawk, and Jet Provost – aircraft on which we practised instrument procedures and low level navigation, formation flying, simulated air-to-air combat, and stalls, spins, and aerobatics, so as to keep ourselves up to speed. Although we worked as test aircrew at the time, we also tried to keep our operational flying skills reasonably honed, so as not to require too much refresher training should a recall to front-line service ever become necessary.

Other aircraft appeared at Boscombe from time to time which added to the variety of the tasks placed upon us. For example, my colleagues and I got to evaluate a captured Pucará retrieved from Port Stanley where it had been abandoned when the Argentine forces capitulated. Our interest in the aircraft was mainly how effective it would be in evading attack from UK jet fighters, and how, by forcing a fly-through in an evasive manoeuvre, the Pucará might turn the tables on its attacker. The aircraft was fitted with two Hispano 20mm cannon and four Browning machine guns, and so could do some damage if it got its sights on another aircraft, however briefly. I also flew ETPS's two-seat Lightning, the Dornier Fantrainer, the prototype Tucano, the ill-fated Optica, and the venerable Chipmunk and Harvard. Flights of the Fantrainer, Optica, and Tucano were for evaluation purposes, but flights in any aircraft, old or new, represented valuable experience (and sometimes quite a lot of challenge too). Lest you should think that all we did was fly, I should perhaps add that most trials and evaluation flights required detailed written reporting, which recorded our results, recommendations and conclusions – and these would contribute to the final report compiled by the trials officers and divisions at the end of any trials series. It would be these final reports that would represent the Establishment's recommendations to the Service chiefs regarding the clearance (or otherwise) of any new aircraft or equipment before entering operational service. I also couldn't avoid the inevitable pile of administrative paperwork that any CO has to attend to – and an in-tray that my PA, Edwina ('Ted') Parsons, regarded as her sacred duty to keep full.

Crashing the Scout

Looking through my logbook, my flying as OC 'A' Squadron was wonderfully varied and interesting, but it was my eagerness to fly anything and everything that nearly became my undoing. On 28 September 1982, the CO of the rotary-wing aircraft test squadron, Commander Geoff Cryer RN, invited me to accompany him on a routine air test of Scout XT647, which had recently undergone major servicing. I had had limited exposure to helicopters at the time but had always enjoyed the experience and so was quick to accept the invitation. It was a rather overcast day as I remember, with a gentle wind blowing from the south. After start, therefore, Geoff hover-taxied the aircraft from 'D' Squadron's aircraft dispersal area to the adjacent southerly grass runway and called for take-off clearance.

Accelerating from the hover to climb speed in a helicopter is not very different to the technique required for the Harrier: engine power is increased and the nose is lowered a little to direct some of the vertical lift or thrust aft (from the rotor blades or engine nozzles respectively). In the Harrier, thrust is increased by moving the throttle forward; in a helicopter it is achieved by raising the collective lever. All this Geoff did in his Scout, but his nose-lowering was rather more dramatic than I had expected, and the nose-down attitude that resulted, rather more extreme. It briefly crossed my mind that Geoff was trying to impress his fast-jet-pilot passenger by showing me a fighter-pilot-like transition, and I glanced across at him to show that I was duly impressed. As soon as I saw the intent look upon his face, however, I knew that something was wrong. From the strain apparent in his right arm and the clench of his fist, I could see that he was struggling with the cyclic control, which for a light and responsive aircraft like the Scout, appeared surprisingly rigid.

Meanwhile, the aircraft's attitude had remained at about twenty degrees nose-down and the aircraft was consequently accelerating very fast. Geoff held his collective lever at the top of its travel, demanding as much power from the engine as he could get, but even as the airspeed continued to increase, he seemed unable to pull the nose up. Worryingly, a slow but seemingly unstoppable roll to the right was also developing, which turned our initial climb into a descent. Geoff had put me in the right-hand seat, and so as the roll continued, my view through the transparencies on my side of the cockpit became a green blur as the runway's grass surface began to race towards me. We had reached a maximum height during the first stage of our transition of perhaps 100ft, but now, as the roll continued, we began to plummet earthwards at an extreme bank angle.

Our airspeed must have reached about a hundred miles per hour when the right skid hit the soft ground and dug in. With this abrupt arrival, the Scout's forward momentum was translated into a rapid rolling motion. In the next few moments, the leading rotor-blade hit the ground hard, ploughing a deep furrow. With so much kinetic energy to dissipate, the aircraft at once began to cartwheel, turning head over heel. Being on the right-hand side of the cockpit, I was to get the best view of the ensuing triple horizontal axel as we flipped over and over. This inelegant manoeuvre was accompanied by the crashing and screeching of tearing metal and shattering plexi-glass. With each succeeding bounce, different parts of the aircraft took the impact. In turn, all the rotor blades buckled, the windscreen and side transparencies imploded, the doors were ripped from their hinges, and finally the tail boom bent in two and doubled back upon itself. Meanwhile, we two pilots enjoyed the most extreme fair-ground ride that you can imagine. But there was no time to contemplate our fate. It was all over in seconds.

We ended our short flight sitting in a stripped-down skeleton of tangled metal, with a long trail of wreckage littering the grass behind us. The remnant of our battered Scout had fortunately finished its journey top-side up rather than up-side down, and suddenly all was still. The high-pitched whining noise in my ears, however, told me that the Rolls Royce Nimbus engine was still running. I was instantly alive to the possibility of fire. It was a wonder indeed that the fuel tanks had not ruptured in the rough and tumble. If they had, the inevitable cloud of atomised AVTUR could so easily have become a fire-ball, which would probably have been the end of us. Spurred into action by this frightening thought, I grabbed the only means that I could see of shutting down the engine – the low-pressure fuel cock by my left thigh – and pulled it sharply closed. Amazingly, we both walked away from the heap of tortured metal with only minor cuts and bruises.

Geoff's squadron aircrew told us that they had watched the whole incident from their crew-room window and were as astonished as we were to see us walk away unscathed. During the subsequent accident investigation, it was established that the loss of control was due to the misalignment of a valve in the starboard cyclic control system. It was the sheer agricultural ruggedness of the Scout's airframe that saved us – and the robustness of its fuel tanks.

Chapter 17

I received my next posting at the end of September 1983, having completed two-and-a-half very fulfilling years as 'A' Squadron's CO. This new post was initially a four-month secondment to the RAF operational training directorate at Adastral House, Holborn, before taking over as Assistant Director, Flying MoD (Procurement Executive) in the St Giles' Court building in Bloomsbury in London's West End. Both these posts meant more of the hated commuting and thus yet more time away from home. I had endured daily commuting from deepest Wiltshire during my first MoD tour in Whitehall and it had worn me out, so this time I decided to commute on a weekly basis, arriving in my office on Monday mornings, living in town during the week, and returning home on Friday evenings. As compromises went, this was at least more physically bearable than daily commuting with its four hours of travelling each day; but being away from home four nights every week was not very satisfactory from a family point of view. For the first four months, I took over the garret bedsit of a mews townhouse near Marble Arch from the wing commander I had temporarily replaced (he had been sent to the Falkland Islands on a four-month unaccompanied tour). This tiny room was on the top floor of the London home of Peter Rees MP, the then Chief Secretary to the Treasury, with whom I shared the breakfast table every morning, and only thirty minutes away by bus from Adastral House. This was very convenient, but when I moved to my new post at St Giles' Court, some arcane RAF regulation on eligibility criteria for central-London living-allowances forced me to move to the outer (and thus cheaper) reaches of town. I was allocated a room in the officers' mess of RAF Bentley Priory, the old wartime Fighter Command headquarters in Stanmore, a lovely former nineteenth-century stately home set in beautiful grounds. It would have been a most desirable residence for me but for the long trek to and from my office every day, crammed nose-to-nose with other miserable commuters on the Northern Line. Although I was eventually permitted to rent a basement bedsit in South Lambeth – which made my daily commute a little easier – it was clear to me that unaccompanied London living was not a lifestyle that I could bear for long.

My new boss at St Giles' Court was Air Commodore John Wilkinson, who had been the Superintendent of Flying Division at Boscombe Down in the early 1970s during my first tour there. As MoD staff tours went, this was an interesting one with lots of autonomy, especially since I spent most of my time on the road or in the air visiting test bases and trials support facilities around the UK. My number two was Squadron Leader Dennis Stangroom, a former Boscombe Down test pilot with whom I had worked before. Between us, we ran a small department of retired RAF officers and civil servants who managed the MoD (PE) aircraft fleet and and its aircrew.

The MoD Procurement Executive was responsible for research, development, production and purchase of weapon systems and equipment for all of the armed services. We, in the PE Flying directorate, managed a fluctuating fleet of over 200 trials or support aircraft at any one time, comprising around forty different aircraft types. These aircraft were divided between A&AEE at Boscombe Down (including those of ETPS), the Royal Aerospace Establishment airfields at Farnborough, Bedford and other out-stations, and the aircraft manufacturer's test airfields, such as Warton in Lancashire and Dunsfold in Surrey. One of our responsibilities was to perform an annual inspection of flying operations at all those bases to ensure that all the MoD (PE) safety standards and procedures were being followed. In truth, the unit COs and company chief test pilots were such highly professional managers that they didn't need us breathing down their necks. But annual checks and audits were well embedded as healthy practice in an aviator's psychology, and no one thought themself so perfect that they could not learn more. When errors were made, moreover, the military culture was to admit, discuss, and promulgate the lessons learned from mistakes – to air them in open forum – not to hide them, but to spread the word so that others might avoid the same fate – not to blame and shame, but to applaud honesty and honour in having the courage to own up. In this culture, it was easy to be a regulator, and our inspections were, like those of the CFS Examining Wing mentioned earlier, collegiate rather than confrontational affairs. Later in my professional life, as a college vice-principal and a NHS hospital director, I saw that there was much that the education and health services could learn from us aviators in these respects.

Return to Boscombe Down

I spent a year as AD Flying (PE) before being promoted to group captain and receiving my orders to report back to Boscombe Down, this time as Superintendent of its Test Flying and Training Division. It was during that year at St Giles' Court that I had started to have conversations with the RAF personnel department at Barwood in Gloucester about the future course of my career, hoping for a change of scene. With three test flying tours already under my belt by that time, I was beginning to think that I had done enough of that and wanted to try my hand at something different. Besides, remaining in the familiar field of flight testing, I argued, even at the higher rank of group captain, would feel a bit repetitive. At 40 years of age, I was then still young enough to crave change for its own sake and had suggested, rather than another tour in MOD (PE), that the command of a suitable RAF flying station might give me the breath of fresh air that I felt I needed. If not a front-line base, then a flying training base would surely be entirely suitable for someone with my earlier background as an A1 Central Flying School flying instructor – perhaps a basic or advanced flying-training airbase, or even an operational base where training took place, like the Tri-national Tornado Training Establishment at Cottesmore. But it seems that the die had been cast. For better or worse, I had made my name in flight-testing and this was where I was destined to stay.

After three months of refresher flying on Hawk in Anglesey and Gazelle in Shropshire, I took over from David Bywater, the outgoing Superintendent of Flying Division at Boscombe Down in April 1985. I inherited his large married-quarter on the base (in which neither I nor Caroline really wanted to live), and also his office with a bay window and balcony that overlooked the airfield. My bailiwick included the management of the airfield and supervision of the three flight-test squadrons, the Empire Test Pilots School, air traffic control, the meteorological office, and the station fire service. The view from my office would have been coveted by any aircraft enthusiast, but the job was mainly administrative and supervisory, with some continuity flying on Hawk, Harvard, and Gazelle to keep my hand in – and sometimes the opportunity to fly other aircraft as co-pilot on suitable test flights. I suppose that this change of role was to be expected as I climbed the professional ladder, but as I gazed enviously out of my panoramic window at all the aircraft comings and goings on the vast concrete acreage stretching out before me, I began to wonder what the hell I was doing there. My wing commanders, John Bolton, Colin Cruikshanks and Dennis Stangroom

(COs of ETPS, A and B Squadrons), and Commander Simon Thornewill and Lieutenant Colonel Wally Steward (successive COs of D Squadron) were thoroughly professional, experienced test pilots and they did not need a lot of supervision from me. The last thing that I had wanted when I had been in their position was the heavy breath of my group captain ventilating my neck, and so I tried to steer my division with a light touch, playing my part as it was written – approving the trials flying instructions for my test aircrew, encouraging, recommending and promoting, advising when advice was called for, resolving issues when they needed resolving – all pretty standard stuff for a boss – just not very exciting. Needless to say, I threw myself into the job with an obsessional desire to do my best – for my squadron commanders, for my heads of service, and for my staff – and to make things better and more effective at A&AEE in whatever way I could. But at heart, I knew that the most exciting years of my RAF career were already behind me, and while my air force future might hold more promotion if I remained in the service, there was no post that I could identify ahead that raised my pulse rate.

In an attempt to stave off this gloomy feeling that the end of my aviation career was nigh, I took every opportunity to fly with my squadrons whenever I could. I flew simulated combat and formation-flying sorties against 'A' Squadron's pilots (and was regularly humiliated by Rod Fredriksen, freshly back from his Falkland Islands combat success); I leapt out of a Hercules from 12,000ft on a so-called accelerated free-fall descent onto Salisbury Plain, and also joined John Bolton on an ETPS static-line jump into Studland Bay; I flew as co-pilot to Dennis Stangroom in 'B' Squadron's Canopus, the A&AEE De Havilland Comet, to Thule in northern Greenland, and then on to the North Pole for navigation trials; I was a Tristar co-pilot for flight-refuelling trials, including refuelling from VC10 and (famously) the Buccaneer; and I flew with 'D' Squadron's Chris Pittaway on several Sea King helicopter trials, including deck landings on HMS *Brave* and on RFA *Gold Rover* in Lyme Bay at night. I was also air-lifted on and off HMS *Endurance*, the Antarctic Ice Patrol Vessel, at the invitation of an old National Defence College friend, Captain Tom Sunter RN, aboard his new command. And finally, as titular head of the Empire Test Pilots' School, I played a sort of figurehead role at ETPS events from time to time – presiding at annual McKenna graduation dinners, accompanying the School on some of its overseas visits, and flying the USNTPS F-18 Hornet on my last-ever fast-jet flight.

But while all this was interesting and often fun, the creation of an annual ball for all Establishment staff, not just the usual hierarchy, was perhaps one of my more worthwhile projects. Working with a small team of engineers and technicians, headed by Gerry Smythe, my old engineering line-chief from 'A' Squadron, we created a 1940s-style annual dinner-dance in the ETPS hangar, in which we draped parachutes and red, white, and blue rosettes, and filled with the music of a Glen Miller tribute band. We danced under the wings of aircraft of the era, dined on trestle tables with gingham cloths and candles, and did our best to dress the part. We raised thousands of pounds for local charities, but perhaps more importantly, we helped to lift spirits – our primary objective – at a time of some uncertainty and unrest in the Establishment with substantial change looming ahead. With all these extra-curricular activities, things were not so bad.

Changes in the wind

As Superintendent of Flying, I had become a member of the 'Forum', the Establishment's senior management board, chaired by the commandant and chief superintendent (chief scientist) and the five divisional superintendents – three senior civil-service engineers or scientists and two group captains – me included. Air Commodore Graham Williams was the commandant and my direct boss for most of my tour. A former Hunter pilot in Aden during my time there and a distinguished test pilot himself, he was a good man to work for. Like me, I suspect he would also rather have still been actively test flying than stuck most of the time in his office.

I was elected a Fellow of the Royal Aeronautical Society in 1987 and became a member of one of its research and development monitoring panels shortly thereafter. From this new perspective, I saw that things in aircraft development and testing were changing fast and that we at Boscombe would have to change the way we did things too. In recent years, the amount of test flying had been reducing year by year with the reducing number of new aircraft and systems coming into service. With less test flying to do, more continuity flying was required to keep our aircrew at a safe level of currency, and the cost of this was becoming disproportionately high. The nature of our work was changing too. Acceptance testing was diminishing in volume as the number of new military projects decreased, and more and more of the operational testing and evaluation of weapon systems was taking place on designated squadrons or units of the mainstream RAF. This trend was set to continue, and this meant that the recruitment, training and deployment of qualified test aircrew in the future would have to reflect these changes too.

CHAPTER 17

I soon found myself under pressure from the MoD to reduce the number of test aircrew at Boscombe and combine the fast-jet and heavy-aircraft test squadrons under one CO rather than two, to make more efficient use of resources. Other testing organizations in the UK and abroad were also cutting back and amalgamating for the same reasons, and it was clear that the training offered by ETPS would also have to be reviewed.

On the one-year ETPS course, every participant was trained for every type of test flying role (in separate fixed or rotary-wing streams) regardless of what future employment lay ahead. Fixed-wing test-pilot students in particular would fly fast-jets as well as transports on test exercises ranging from performance and handling testing to spinning and advanced weapon systems and avionic trials. Yet it was clear that not all client organizations sought all of this. Many had more specific needs and would not want to pay for training that was unnecessary, nor want to lose their aircrew from productive service for a whole year. It seemed to me, therefore, that a more versatile approach would be needed in future, including the creation of specialised short courses that would be more economic for our clients in cost and time – courses that could be completed for particular testing roles as the need arose. These shorter, more focused courses, I proposed, would be built on suitable training modules from the long course, and constructed to meet a range of client needs – in commercial as well as military aviation. The US-based National Test Pilot School, a commercial training provider based in Mojave, California, was already making its name with this approach, and I thought that we could learn from this. I had a battle on my hands with the established mindset of the day, but financial pressures and the winds of change would eventually prove irresistible.

I spent three years as Superintendent of Flying, but as the end of my tour approached, I began to think again about my future in the RAF. To me, two things seemed fairly certain: first, I had had my best flying years, and whatever might lie ahead in the RAF was likely to be dull by comparison; and second, the great majority of the ten years of service remaining before I reached the normal retirement age (of 55) would inevitably be spent behind a desk, most probably in London, the idea of which I had come to loathe. Wherever my path might lead me if I remained in the service, moreover, I was certain that I would face more years of living away from home or spend more tiresome, lost hours in long-distance commuting. Caroline had had enough of it too. She had suffered my too-frequent absences and had dutifully upped-sticks and moved house and home with me ten times during my career so far – within the UK and across three continents. I could not expect her to do more. The uncertainty of not knowing when or where we might be sent had been stimulating, even exciting, in our younger lives, but that was before we had put down roots and while our children were still young. Now, the thought of not having control of our future was unsettling. If I had to sit behind a desk, I reasoned, it should be I and not the RAF's Personnel Branch at Barnwood that would choose which desk it would be and what sort of paperwork would come across it. All this considered (carefully and with some considerable regret), I decided to leave the RAF at the end of my tour and try my luck elsewhere.

As a teenager eager for variety, travel, and adventure, I had set myself a lifetime target of doing as many different things as I could and trying out as many interesting activities as might be possible (as well as legal and not too injurious to health!). I had done pretty well to date in putting this adolescent resolve into practice, and I had now come to the point where it was time to move on.

Chapter 18

I was coming up to 45 years-old when I left Boscombe Down for the last time in July 1988 after just over twenty-six years of RAF service. This was classic middle-age-crisis territory. I had a lot of optimism and hope for a new future but no clear picture in my mind as to what it would look like.

Having put service life behind me once and for all, all I really understood about myself was that first I wanted to create a more stable domestic life; second, I wanted something fulfilling to do; and third, I wanted a refreshing change. I also expected to meet all of those criteria while still earning a reasonable living wage.

With these primary objectives, I realised that I had set myself a stiff challenge, and I have to admit that it was at first a little daunting. Over the years preceding my decision, I had toyed with the idea of doing a post-graduate year at university, either an MBA or some other course to help my transition, and the idea of a year out to retrain and reorientate still had some appeal. Already in the spring of 1988, prior to my departure from Boscombe, I had successfully applied for such a course at Brunel University starting in the September of that year. At that time, I regarded it as a fall-back option – something to have up my sleeve in case nothing appropriate came my way – but it would prove to be just the thing I needed, for it would eventually lead me in a completely different direction.

Even before I was due to graduate from Brunel, however, I was already scanning the appointments pages of the broadsheets and aviation press just in case something suited to my aptitude and interests came up. By chance, I spotted a recruiting advert for flying instructional posts at the Oxford Air Training School at Kidlington, a former RAF airfield located between Oxford and Bicester, where I had started my RAF flying instructional career at OUAS twenty-two years before. It was a financial lifeline that would do for the present while I settled into a new civilian way of life. The job came with a half-way decent salary and included the necessary commercial pilot and commercial instructor conversion and refresher training that I would need. Eager to clutch at any straw that offered financial rescue, I applied at once.

After all my years of military flying and study and as a former QFI and instrument rating examiner, I thought that passing the theory for my commercial licence would be relatively straightforward. But the ten weeks of ground school required for a dozen or so CAA technical examination papers needed a lot more homework than I was expecting, even though I had studied (and indeed taught) most of it before. The civil instrument rating proved not to be a walk-over either. Carrying out a cross-wind holding pattern and single-engine approach based only on radio direction and range information, needed the mental agility of a circus performer juggling balls while balancing on a tightrope! On my first attempt, I even had a slight tail-wind on the glide-path, which made things just a bit tricky.

Still, I got my commercial tickets and joined the instructing staff at Oxford ATS, training British Airways, Monarch Airways, and Aer Lingus cadet pilots to fly. It felt as if I had put the clock back to my favourite days at the Central Flying School at Little Rissington, simply having exchanged the four air force blue stripes on my epaulettes for four gold ones. And instead of Chipmunks, I was now flying the Piper Cadet and Warrior, the Slingsby T67, and the twin-engine Piper Seneca, teaching students from ab-initio to solo, and from aerobatics to advanced instrument and procedural flying. It was such a relief to be back in the air and once more the master of my domain, teaching highly motivated young men and women who hung on my worldly words of aviation wisdom as if they had been brought down from Mount Sinai itself!

Time was money in the commercial world, and the airlines wanted their students qualified and moved on into the next stage of training (line-training on airline types) as quickly as possible. Indeed, our contracts with the airlines imposed additional costs on the School if graduation was delayed. Every unrecoverable day lost through bad weather, therefore, was not only frustrating, but could also hit the school's bottom line. To mitigate this risk, the school took the decision to form a flying-training unit at Lakeland airfield in Florida, then the site of one of the Piper Aircraft assembly plants, where clear skies and sunshine could almost be guaranteed. Those of us who wanted to take their student cohort to the USA were soon being encouraged to do so, with accommodation and subsistence provided. Well, of course, I couldn't resist the opportunity for another overseas detachment and was quick to sign up for a six-week stint. Between 10 April and 25 May 1990,

I flew 120 instructional sorties and brought all of my basic students up to instrument flying standard before we returned to the UK for the summer. With an American commercial pilots' licence now in my pocket, I rather fancied a few more of these short detachments, but when I got back, another interesting aviation opportunity dropped into my lap that was to set me on another path altogether.

The International Test Pilots' School was looking for a Principal and Director of Training, a new post for this new and growing enterprise. James Giles, its founder and MD and a former test pilot tutor at ETPS, approached me to sound out my availability. The School was based at Cranfield in Bedfordshire. A former RAF airfield with an 1,800-metre tarmac runway, Cranfield was the site of the then Institute of Technology (now Cranfield University), which incorporated the College of Aeronautics and a range of aeronautical research facilities. ITPS had equipped itself with a training fleet comprising a Pilatus PC-7 turbo-trainer, a PC-6 Porter, a Westland Aérospatiale Gazelle helicopter, and a Gyroflug SC-01 Speed Canard, and these were supplemented from time to time with other aircraft on short-term hire, such as the Slingsby T67 and the Calspan variable-stability Learjet. Arrangements had also been made with foreign governmental clients for flight-time to be made available on national front-line fast-jet aircraft such as the F-16. The company was in its second full year of commercial operations – staffed by a dozen former military test pilots, academics, and flight test engineers working as tutors – and was already attracting students from some of the smaller foreign air forces and aircraft manufacturers. This new business had been won largely by offering training equivalent to that received by students at ETPS but at a much lower price, achieved through lower overheads and direct operating costs.

Being a relatively new business entity, however, ITPS had not yet built up a regular client base, and so its income stream was uncertain. Without a strategic reserve to weather the likely volatility of training revenue, moreover, cash-flow was still a critical problem. My primary focus in managing and developing the School would therefore be two-fold: to broaden the range of courses to suit a wider field of potential clients internationally; and to increase the Company's order book significantly to provide a reliable and regular revenue stream. It was an exciting prospect and an offer that I couldn't refuse.

ITPS's flag-ship course was its eleven-month 'Long' course, which had a syllabus and structure similar to that of ETPS. It was this 'Long' course that to date had brought in almost all of the School's revenue, and this mostly from clients not traditionally those who would have chosen or been eligible to send students to the military test pilot schools. It seemed that the relatively lower price and wider accessibility of the ITPS course had created new demand. This was especially true of the aircraft manufacturing and commercial airline sector. The main problem with this sector, however, was that, by and large, such clients, careful of their bottom-line, were late in booking course places and also late in parting with their cash. Student numbers for each new 'Long' course and consequently the revenue for the year ahead, therefore, were never really certain until just before the start date. With staff contracts to be renewed and aircraft leases needing to be extended (or modified) on the basis of student numbers and expected revenue, the approach of each new school year was bound to be a nervous time.

To mitigate these end-of-year jitters and bring in some extra revenue more evenly through the year, a new activity group had been formed within the School to offer flight-test services on a contract-hire basis. It was hoped that this service, marketed under the name International Flight Test (IFT), would be the means by which staff expertise could be kept current and put to good (income-earning) use when the training workload allowed. Sub-contracting under this banner was already generating some additional cash, but it was marginal revenue that was not contributing to the Company's overheads nor even covering its costs. Nevertheless, these were still early days for IFT, and growth was hoped for. It was also argued that IFT activity kept tutors in touch with real-world research and development, thereby adding to their stature and credentials as tutors. This was undoubtedly true, but there was not much genuinely profitable IFT work to be found at the time, and the time and effort spent pursuing it was proving a bit of a distraction at a critical time in the School's consolidation as a training provider. This was also at a time when there were major shifts underway in the worldwide aerospace training market. As a consequence of the recent collapse of the Warsaw Pact, spending reviews were taking place in many countries that were looking to cut defence costs – and thus realise some of the so called 'peace dividend'. ITPS was thus just in the right position to draw the interest of clients looking for more cost-effective and flexible training options, but this would require some refocusing of the School and some very active marketing. As I took over as the School's director, I knew that this was where ITPS needed to concentrate.

My brief was to put the School on a firm financial footing by broadening its training provision and building up its business. For me, it would be a two-year reconstruction and redevelopment project, and a useful transition into the business management world where I would eventually end up. ETPS' reluctance to move away from its traditional model of a single annual long course had left a gap in the market for ITPS to lead the way with a more flexible approach. We did this by creating a range of new specialist short courses that could be offered throughout the year; courses with titles such as:

> Stability, control, and handling qualities testing;
> Performance testing;
> Avionic-systems testing;
> Simulator evaluation;
> Civil certification and procurement;
> Flight-test instrumentation and data analysis.

We also offered bespoke training packages that were tailored to meet specific clients' requirements, as well as pre-course distance-learning packages, which were offered as an alternative to classroom-based tuition. This remote tutorial provision recognised some clients' desires to reduce the time spent by employees away from their home base (and the associated accommodation and subsistence costs). Students from SAAB, for example, were able to reduce the long course duration from eleven months to eight in this manner.

The new short courses were not intended to replace the standard eleven-month course, for which there was still some demand, but were run in parallel as alternative options geared to meeting client needs. The big financial benefit of these shorter courses for many clients was that they would only pay for the training they wanted when and as they needed it. Attendees at such courses would be tutored specifically in up-to-date methods using the latest technology at the appropriate level for the particular trials they were about to undertake, and sometimes at their own bases in their own aircraft. Extension training, upgrades, and refresher training modules would be made available when required. Skills and techniques learned would thus always be relevant and current, and not have grown stale through lack of use.

The big advantage for ITPS in adopting this approach was that suitable training modules, as well as staffing, could be 'borrowed' from long course resources at little extra cost. This not only increased total revenue but also smoothed the cash-flow, removing some of the 'cliff-edge' year-end jitters previously mentioned. With overheads effectively already covered in our business model, and with the associated direct operating costs of short courses relatively small, the overall profitability of the School's activity increased significantly. With this new portfolio of courses developed, packaged, priced, and marketed, my next task was to attract interest and build sales.

My first initiative in this respect was to embark on a concerted campaign of direct approaches to key individuals in selected foreign ministries of defence, the airlines, and principal aircraft manufacturers around the world. These key decision-makers were identified with the help of the defence and commercial attaches and export trade specialists in UK embassies in our target countries. I received about a 25 per cent response rate from this approach, but as soon as any interest was shown, I immediately followed it up with an offer to visit to discuss the client's training needs further. Once my offer of a visit had been accepted, I re-enlisted the help of the UK embassy attachés and commercial specialists to arrange presentations to other relevant organizations in the country or in countries nearby. In my first two years, I must have undertaken a dozen sales tours to a number of regions around the world, including North and South America, South Africa, the Far East, and throughout Europe. I have to admit that all this travelling became quite a chore, but such personal face-to-face contact with potential customers paid off handsomely. It was on one of my several trips to South America that I had met and befriended Colonel Juan Carlos Sapolski, the chief test pilot of the Argentinian Air Force mentioned earlier. On another, I flew with the Chilean Air Force from Santiago and contracted Chilean students for our courses the following year. Students from Germany, South Africa, Switzerland, South Korea, Australia, Norway, and Spain were also enrolled in the same way. We won the Queen's Award for Export Achievement in 1992 for an outstanding two-year growth record in overseas sales, and it was this growth spurt that brought in significant income at a critical time in the Company's development.

On the basis of this improved financial position and a healthy order-book, we were able to move out of our cramped temporary buildings into a newly refurbished and much more spacious former RAF 'H'-block in the heart of the campus. We also commissioned the adaptation of a Beech King Air E-90 to create an avionics-systems training aircraft by incorporating a digital data-bus, satellite and inertial navigation systems, mapping radar, and a digital analysis facility. In addition, our variable-stability simulator and in-house software for flight and avionics-systems testing had reached an advanced stage of development, and was soon enhancing the syllabuses of new and existing courses. Latterly, I also negotiated flight-time for our students on the RAF's Hawk advanced jet-training aircraft from RAF Valley, which was an important step towards levelling up our aircraft inventory with that of our Boscombe Down competitor. From a small start-up position, the School was fast establishing itself as a substantial presence in international flight-test training. It provided training equivalent to ETPS yet was also more nimble, responsive to demand, and cost-effective. In my second and final year at ITPS, the School was already attracting students from sixteen countries and building up a solid reputation around the world.

With this achieved within the time-scale I had set for myself, I decided it was time to take my leave of ITPS to pursue the other avenues of employment I'd had in mind when I retired from RAF service. The school eventually moved to London, Ontario, as ITPS (Canada) Ltd in 2001, where it thrives today under new management, but I am grateful to James for the opportunity he gave me to contribute to its success in the early years of his project. Moreover, I pay tribute to him for his courage in striking out on his own with such an imaginative and innovative venture. He, I, and the School enjoyed some dynamic, very successful and satisfying years. At the same time, I learned one heck of a lot about financial management and the cut and thrust of running a business – which was to prove invaluable experience in the very different phases of the career which had yet to unfold before me (but not the subject of this book!).

Chapter 19

These days, when I am asked what I did for a living, I might offer a few clipped sentences about one or other of the various business management appointments that I have held in more recent years. But because I am an inveterate entertainer and highly sensitive to the glazing-over of my listener's eyes, I might toss in that I used to be a test pilot, just to enliven the conversation (and make it more interesting for me too). If this little nugget of discourse sparks any interest at all, the response is often something like: 'oh, wasn't that a bit dangerous?' To which I might reply flippantly, 'not a lot more dangerous than riding my bike', which I do a lot. This is probably not true on the basis of accident rate per head of the relevant participating population, but it gets a laugh when I wryly tell them how many times my cycling friend, Steve, has come off his bike and finished up in plaster. I wouldn't anyway describe the work of a test pilot as more risky than the work of any military pilot flying fast-jets in any role. While the test pilot might encounter situations for which the rule book has not yet been written, pilots on front-line squadrons, even in peacetime, face risks that are just as great and often greater. This is because in trying to simulate wartime conditions in military training – so as to be fully prepared when the balloon really does go up – things are often made more demanding than the real thing, which sometimes magnifies the risk out of proportion with the training gain. For example, since the Second World War, more accidents and fatalities per flying hour have occurred in training than have occurred in operations. In a report of the fiscal year 2019, members of the US Armed Service Committee said that nearly four times as many military personnel died in training or non-operational accidents as were killed in combat.

Most flying fatalities since the Second World War have had nothing to do with flight testing or being shot out of the sky. In my thirty-year flying career, a dozen pilots within my own close circle have been killed in aircraft accidents. There aren't many peacetime professions where you could lose so many friends and colleagues. Only one of these was directly the result of a serious technical problem – a critical system failure, made worse by a string of technical malfunctions or maintenance issues like the ones described in earlier chapters. In all of the remaining cases, these good men tragically lost their lives because of simple errors or misjudgements – not technical malfunctions or catastrophic engine failures like mine when there was time to eject. This is not to say that those pilots were culpable or negligent. Only one of those deaths was due to downright recklessness, and he unforgivably took others with him to the grave. The remainder of those fatalities resulted from what I would call 'honest' mistakes, the sort of human error or aberration that we are all capable of committing daily – and mostly getting away with: slips of concentration; an incorrect instrument setting; some small but necessary action in the check-list or procedure missed; an oversight when working under pressure; a routine rushed when short of time. Simple, single errors like these are often compounded by other contributory technical, environmental, or human-factors elements that make the outcome more serious than the initial error might warrant – an unfortunate conspiracy of circumstances.

An example from my own Hunter days in Aden might be salutary: a case where, but for a bit of luck, conspiring factors might well have led to two fatalities – one of them my own. This was at a time when the Strike Wing pilots of Nos 43 and 8 Squadrons were rostered two at a time for weekend QRA (quick reaction alert) – all geared up and ready to scramble at ten minutes notice should air-support be required up country. On the day in question, I was the pair's leader on standby, sitting in the ops room idly flicking through a flight-safety magazine, when the alert came. My number two and I were quickly briefed on our mission by the Wing's Army Liaison Officer. It appeared that a commando unit had got itself pinned down in an ambush in the Radfan mountains eighty miles or so to the north, and had requested urgent air support. Our two standby Hunters were fully fuelled and armed, each with eight 60-lb HE rockets on the outboard wing pylons and a full magazine of 30mm HE rounds for each of the aircraft's four Aden cannon. We marked the coordinates on our maps and hastily planned a route. Rockets added a significant amount of drag, and the higher fuel consumption meant that our time in a target area would be relatively short – perhaps only ten minutes before we would have to return.

Feeling the pressure of time, we ran out to our waiting aircraft, started up, and taxied out to the runway in some haste. This sense of urgency was the first factor in the chain of events about to unfold. With runway 08 in use, there was a good mile to taxi before we reached the take-off point, but just as I was about to call for take-off clearance, we were recalled. It appeared that someone higher up the chain of command had decided that our weapon load should be increased to sixteen rockets on each aircraft rather than the eight already loaded. It was a mile to taxi back for the additional rockets to be uploaded while we sat sweating in our cockpits with our engines running, and then another mile back to the take-off point. This used up precious fuel and time, which added further pressure. Our quick reaction to the Army's urgent call would thus not be quite so quick as we had hoped! In an attempt to make up lost time, I increased our cruise airspeed above that which I had planned. Since drag is a function of the square of the airspeed, this dash into the mountains would burn significantly more fuel per mile than I had originally estimated. Malign factors were already beginning to add up. The delay had depleted our fuel, and now we were burning yet more. Moreover, with no opportunity to check the operations manual while we had sat briefly in the chocks, I had not recalculated the higher fuel usage in the sixteen-rocket configuration, which was heavier and more draggy than the one I had planned for. Not only would this higher fuel consumption reduce our time available in the target area, but we would also need to allow more fuel for our return to base if any of our rockets remained unused. This was my first error; while the rocket load was being increased, I should have closed down the engine, topped-up with fuel, called for the manuals, and adjusted my original fuel calculations. The pressures of time and the panicky expectations of the ALO had seemed overriding at the time, but a more experienced pilot might have obeyed his instincts and done exactly that.

Yet more malign factors were soon to arise. When we arrived in the target area, the visibility was really poor and there were no identifiable navigation features. The Radfan is a barren, arid mountainous region: a labyrinth of rocky ridges and dried-up wadis where one craggy peak looks much like another. With no navigation systems or aids to guide us, we could not immediately locate the pinned-down unit. We set up a search pattern, but to no avail. We were in good radio contact with the forward air controller by then, but could not identify his location despite his descriptions. I called for smoke to help us find his marker, but in the poor visibility, we could not see this either. It was like flying inside a dirty goldfish bowl. I began to doubt both my navigation and the coordinates we had been given for the unit's location, and so we expanded our search. Again no luck. And all this time, the extra weight and drag of the increased rocket load was depleting our fuel much faster than I had expected.

It did not seem long, after sweeping the area back and forth, scanning the ground vainly for clues, before fuel had reached my calculated 'Joker' fuel level, which should have meant an immediate return to base. But desperate for the mission to succeed, I stayed on perhaps a minute or two longer than I should. This was error number two! At 'Joker' fuel, I should have admitted my failure and departed for home immediately rather than eat into my safety margins. But I didn't, and this error was compounded further because my 'Joker' fuel had not been recalculated for a return to base with sixteen rather than eight unexpended rockets. I had even less margin than I thought.

Finally giving up our search, we headed south for home. On the mountain slopes below us, the commando unit would somehow manage to extricate itself without our help and without mishap. Because fuel was tight, I decided to climb to an altitude where our Avon engines' specific air-range would be better. Even so, it soon became evident that the higher drag was narrowing our already slim fuel margins alarmingly. As if this were not enough pressure to test our mettle, yet more malign factors now arose. As we proceeded southwards, low-level cloud began to form beneath us, broken at first but eventually obscuring the ground completely. The Khormaksar runway lay on an isthmus of flat, low-lying ground, with the eastern and western boundaries of the airfield reaching the shore lines of the Indian Ocean and Aden Bay. A few miles south of the runway, however, Shamsan's extinct volcanic crater rose 1,500ft above sea-level. If we were going to descend into the airfield blind, therefore, we would need help to ensure that we avoided it. I radioed the base to declare our low fuel state and request a priority radar-controlled descent – only to be told that the radar had developed a technical fault and had just been shut down. I now had no reliable position information. The air-traffic controller was able, however, to confirm that we were still to the north of the airfield by using the bearings of our radio transmissions; but he also reported that the cloud base was around 1,500ft. This gave us no margin between the cloud base and those rocky peaks!

Receiving this information, a prudent pilot with more fuel to spare might have flown out to sea and carried out a blind let down on barometric altitude, but those extra minutes that we had spent in the target area had denied me that option. Fuel was now becoming so critical that there was nothing for it but to plunge through the cloud, trusting my instincts (those fabled instincts again!) that we would emerge positioned for an approach to the airfield – and not come too close to Shamsan's volcanic slopes.

I called my number two into close echelon starboard and commenced a left-hand turning descent. The cloud was not thick, and so we quickly emerged into the clear air below, perfectly positioned for a run-in-and-break for runway 08, just as I had hoped. My homing instincts had proved sound, thank goodness. But glancing over my right shoulder as we straightened, I saw that the cloud-base brushed the tops of the volcano's ridge and that our blind flight path must have taken us within a metaphorical hair's breadth of hitting it. My questionable decision had paid off, but had I made it just a few seconds later, it could so easily have been the final link in the chain of contributory causes that led to our demise. In the critical fuel situation that I had allowed to develop, the unexpected formation of the low cloud layer and then the airfield's radar failure had eaten up all my safety margin. I had allowed myself to be backed into a corner. These last two factors were the final links in the chain that left me no option but to take a risk and plunge blindly down through the cloud. My number two and I were fortunate to have got away with it.

In my defence, I was a young and inexperienced fighter-leader then. But from this salutary experience, I grew much older and wiser overnight. I learned not to take anything for granted – to check and check again. I also learned in future to ask myself all those 'what if' questions when planning my flights – what if the weather changes; what if the radar or some item of aircraft equipment fails; what if something else goes wrong – and to consider all my fall-back and escape options so as to build in extra margins for the unforeseen, and even to allow a bit more for the unlikely. My wife often criticises me for arriving at the airport an hour earlier than necessary for a holiday flight, but caution is now deeply ingrained into my psyche – what if I have a puncture; what if there is congestion on the route; what if the airport car-park pick-up coach is late, etc. She finds it exasperating, but I am easier in my mind as we travel. As is often said in military aviation circles: there are bold pilots and there are old pilots, but there are no old, bold pilots.

I have already mentioned my involvement in the MRCA P-05 PIO incident inquiry. I was also involved with the inquiries into three other aircraft accidents either as a member or chair of inquiry boards, or in technical support of legal advocates making claims on behalf of the pilots concerned. Two of these accidents in particular resulted from chains of events like those above, where several factors combined to make them lethal, even if no single factor individually was necessarily so. It was clear in all of the cases, nevertheless, that pilot errors not dissimilar in magnitude to mine were present – errors of judgement, lapses of skill, momentary losses of concentration, simple mistakes – not all of these present in every one of those accidents, but one or two such errors in each. There had been other failures present in those accidents too, even before the pilot got into the cockpit, that contributed to the chain of causes that culminated in a catastrophic end – maintenance failures, supervisory and training issues, pressures of expectation, weaknesses in oversight and guidance, or simply an absence of clear rules or advice. Other people were responsible for these contributory causes, and if any one of those people had performed differently – with more skill, care, wisdom, foresight, clarity etc – the chain might have been broken, the accident averted, and a life saved. The pilot's error is almost inevitably the last in the chain, and while not always the one most culpable, it is the one that is often easiest to blame. Others not as visible may have borne some of the responsibility to a greater or lesser degree, but it is the pilot, crew members, and their families who are ones who pay the price.

There but by the grace of God go all of us.

While my departure from ITPS might have marked the end of professional flying for me, it was not the end of my flying altogether. I was soon to become the chief flying instructor (voluntary, unpaid, and part-time) of the Bustard Flying Club at Boscombe Down, whose members were largely current and former MoD employees. I would also contribute to a number of light aircraft and cockpit design projects in the years to come, two of them happily with my son, Andrew, himself a former RAF pilot. The most notable of these, the ZIG certified light-aircraft design project funded by the UK's Department of Trade and Industry, was particularly interesting from the design point of view.

In my days as a military test pilot working on Tornado in the 1970s, I had to walk through an acre of draughtsmen's drawing boards every time I visited the British Aircraft Corporation's Tornado design office. Hundreds of draughtsmen and women were employed and thousands of working drawings were produced – all hard copy – for use by a huge workforce of skilled craftsmen and women employed in the manufacture and assembly of individual components. The technologies of computer-aided design, modelling, and manufacture was then in its relative infancy, and so much of the designing needed to be done manually with rulers, set squares, and slide-rules. The ZIG project of the 1990s, by contrast, employed the very latest computer-aided techniques available at the time for design, aerodynamic modelling, and eventually in robotising manufacture.

ZIG was a much smaller project, of course, but pro-rata, in terms of design hours per widget, it required a fraction of the time, and its manufacture would have been largely automated. I was the project's so-called 'chief test pilot', a role I undertook purely out of interest as a part-time (unpaid) member of a young and dynamic design team. Although destined never to fly the aircraft myself, I was extensively involved with market research, with the cockpit design, and in defining the aircraft's specification. What struck me, in contrast to my Tornado experience, was the revolutionary impact of computing software on every aspect of our work, from performance modelling to cockpit and display design. I remember one particular session where, wanting to be sure that our design was optimised for its intended role and market, we experimented by varying the parameters of the aircraft's specification. By changing the performance, engine power, and weight and load-carrying parameters, we could evaluate the pros and cons of trading one against another. With this iterative process, the compromise which best matched the potential market could be established (a process which aircraft design invariably becomes). As we varied each parameter, the shape and configuration of the aircraft 'morphed' almost instantaneously before our eyes on a detailed three-dimensional depiction projected onto an overhead screen. The process helped us quickly to verify our assumptions and settle on a final design. A similar job in the 1970s would probably have taken months.

More recently, as the part-time, unpaid chairman of Aero Safety Systems Ltd, I have been involved with the development of 'smart' devices that monitor safety and system performance in light sports and experimental aircraft, particularly those in the USA. This has included designing and manufacturing a warning annunciator panel for the Hunter T7 now being flown by ITPS (Canada) in its test pilot training programme.

Looking back, I see those thirty years as a professional aviator as some of the best of my life, working with some outstanding aviation professionals – aircrew, scientists, and engineers – with some lasting friendships formed. Once the aviation bug enters the bloodstream, flying becomes addictive, an affliction that requires extensive therapy even if you resolve to kick the habit. Even as I approach my final chapter, written at the grand old age of 77, I am still a frequent flyer and a shareholder of an EV97 Eurostar. We fly our little red monoplane from Thruxton, a former RAF airfield (and motor racing circuit) lying almost directly under the flight path of the main runway at Boscombe Down, the site of some of my fondest memories and former glories. It is an appropriate place to finish up! I am also a volunteer guide at the Boscombe Down Aircraft Collection at nearby Old Sarum airfield (another historic aviation site) near Salisbury in Wiltshire. The

Collection, based in two original First World War hangars filled with aviation exhibits spanning the period 1914 to the end of the Cold War, has more than a score of former Boscombe Down test aircraft in which you are able to sit at the controls and imagine yourself to be the pilot. You (and your children and grandchildren) might want to spend a day there, where my fellow guides and I would be delighted to tell you more of our stories.

Well, this brings me to the end of my tale, or at least that part of it related to aviation. As I said at the beginning, I have been lucky – a helpful faculty that beats most of my other attributes, which is just as well. Certainly I was very lucky as far as my several narrow escapes were concerned and in the fantastic opportunities that were placed before me, but I was also extremely lucky throughout my service career to work with some highly talented and dedicated individuals, most notably those I encountered during my four tours of duty at Boscombe Down. Many have been mentioned in these pages, and all those and other former colleagues are remembered with great respect and affection. In my subsequent twenty-seven years of professional life, I found myself working in some very different fields of employment – in education, in the national health service, and in the criminal justice system. All this was fascinating, challenging and worthwhile work too, but nothing ever quite matched the sense of shared mission, comradeship, and adventure that I enjoyed as a military flyer and test pilot – and it was certainly not as much fun!

Epilogue

Ron Burrows finally made the changes to his life that he had been seeking. Some readers might find it difficult to understand his desire to leave professional aviation having got so far, and to follow his desire to try something different. After all, as he would have said himself, flying certainly beat working for a living! But while he would not have given up a single one of his thirty professional flying years, he had come to recognise that more of the same would have felt like going round the buoy again. He had done so many challenging things, had flown so many different aircraft types in so many different places – indeed, had survived so many emergencies and near misses – that any flying that might have been ahead for him would almost inevitably have felt unsatisfying by comparison.

That post-graduate year that he spent at Brunel University in 1988 and his business development experience at ITPS eventually set him up for another thirty quite different professional years ahead. Until ending gainful employment at 76 years of age, his civilian professional life was as varied as it was unpredictable, determined as much by opportunity as any grand plan.

His first non-aviation role was as the vice-principal and business manager of a city further education college, which, like all FE colleges at the time, became an autonomous incorporated business institution under the 1988 Education Reform Act. Ron managed this transformation, putting in place all of the institution's new business systems and services, replacing those that had previously been provided by the local education authority, and then went on to modernise and develop the College's built estate. He spent twelve years in this role while latterly also becoming a justice of the peace, in which public office he served the Wiltshire magistracy for sixteen years. Retiring from full time employment in 2001, he joined the board of his local National Health Service Hospital Foundation Trust as a non-executive director, serving as such for nine years, eventually becoming its deputy chairman. In 2011, he took on a voluntary role for two years as project director for TradeAid, a vocational-training charity working in southern Tanzania. It was in this role that he was also given the task of project-managing TradeAid's restoration of an historic German colonial building, Livingstone House in Mikindani, the last staging post of the intrepid explorer's final expedition. Ron then returned to the NHS as the founding Chairman of a subsidiary company providing healthcare-linen services across central and southern England. He finally retired in 2019

In his spare time, he qualified as a RYA Coastal Skipper and a PADI Divemaster, has self-published three historical novels, and has ridden his bike the length of the USA and the breadth of Europe.

He still flies as a private pilot.

Index

Squadrons and Units

Places, Bases & Organisations

Aircraft

INDEX

Equipment, Weapons, & Trials Flying Categories